Springer Theses

Recognizing Outstanding Ph.D. Research

For further volumes:
http://www.springer.com/series/8790

Aims and Scope

The series "Springer Theses" brings together a selection of the very best Ph.D. theses from around the world and across the physical sciences. Nominated and endorsed by two recognized specialists, each published volume has been selected for its scientific excellence and the high impact of its contents for the pertinent field of research. For greater accessibility to non-specialists, the published versions include an extended introduction, as well as a foreword by the student's supervisor explaining the special relevance of the work for the field. As a whole, the series will provide a valuable resource both for newcomers to the research fields described, and for other scientists seeking detailed background information on special questions. Finally, it provides an accredited documentation of the valuable contributions made by today's younger generation of scientists.

Theses are accepted into the series by invited nomination only and must fulfill all of the following criteria

- They must be written in good English.
- The topic should fall within the confines of Chemistry, Physics, Earth Sciences, Engineering and related interdisciplinary fields such as Materials, Nanoscience, Chemical Engineering, Complex Systems and Biophysics.
- The work reported in the thesis must represent a significant scientific advance.
- If the thesis includes previously published material, permission to reproduce this must be gained from the respective copyright holder.
- They must have been examined and passed during the 12 months prior to nomination.
- Each thesis should include a foreword by the supervisor outlining the significance of its content.
- The theses should have a clearly defined structure including an introduction accessible to scientists not expert in that particular field.

Christian Schubert

Magnetic Order and Coupling Phenomena

A Study of Magnetic Structure and Magnetization Reversal Processes in Rare-Earth–Transition-Metal Based Alloys and Heterostructures

Doctoral Thesis accepted by
Chemnitz University of Technology, Germany

Springer

Author
Dr. Christian Schubert
Experimental Physics IV
Institute of Physics
University of Augsburg
Augsburg
Germany

Supervisor
Prof. Manfred Albrecht
Experimental Physics IV
Institute of Physics
University of Augsburg
Augsburg
Germany

ISSN 2190-5053 ISSN 2190-5061 (electronic)
ISBN 978-3-319-38389-7 ISBN 978-3-319-07106-0 (eBook)
DOI 10.1007/978-3-319-07106-0
Springer Cham Heidelberg New York Dordrecht London

Softcover reprint of the hardcover 1st edition 2014

Printed on acid-free paper

Springer is part of Springer Science+Business Media (www.springer.com)

Information is a commodity.
It can be traded, sold, and purchased.
And in the end, credits are only as useful
as the secrets they can buy.

—D.B.

Supervisor's Foreword

Since the discovery of a unidirectional magnetic exchange anisotropy in Co/CoO nanoparticles in 1956 by Meiklejohn and Bean [1], causing a shift of the magnetic hysteresis loop, much attention has been paid to the understanding of this so-called exchange-bias effect. But even more attention was triggered by the pioneering work of the Nobel Prize laureates Peter Grünberg and Albert Fert, who discovered the giant magnetoresistance effect, which led to the application in magnetic field sensors using the exchange-bias effect for pinning the magnetic reference layer in spin valve systems. Exchange-bias is most frequently observed in antiferromagnetic/ferromagnetic bilayers due to the interfacial exchange coupling of frozen uncompensated spins [2]. But in recent years a variety of different material systems have been investigated, including ferrimagnetic bilayers [3] employing amorphous rare-earth-transition-metal alloys. The magnetic coupling in such systems consists of two types of pair interactions, an antiparallel magnetic exchange coupling between the rare-earth and transition-metal moments and a parallel magnetic exchange coupling of the transition-metal moments themselves. Taking advantage of the tunability of the exchange interaction between the ferrimagnetic layers, the shift of the hysteresis loop can even be reversed. In combination with a hard magnetic layer, a large exchange-bias effect has also been reported [4].

In this regard, the goal of this thesis was to investigate the magnetic structure and reversal mechanism of single amorphous ferrimagnetic Fe–Tb alloy thin films and ferromagnetic [Co/Pt] multilayers with perpendicular magnetic anisotropy in comparison to exchange coupled heterostructures. With this comprehensive study, Christian Schubert was able to provide a deeper insight into the magnetization reversal process, the magnetic moment configuration, and the role of the interfacial domain wall, occurring in exchange-biased Fe–Tb/[Co/Pt] heterostructures. Furthermore, this study was supported by element-specific X-ray magnetic circular dichroism measurements, revealing the interaction between the moments of the Fe and Tb sublattices as a function of external magnetic field and temperature. The understanding of the peculiar magnetic coupling phenomena in rare-earth-transition-metal alloy-based exchange-bias systems presented in this thesis opens up interesting paths for spintronic applications.

Augsburg, March 2014 — Prof. Manfred Albrecht

References

1. W.H. Meiklejohn, C.P. Bean, Phys. Rev. **102**, 1413 (1956)
2. F. Radu, H. Zabel, Springer Tracts Mod. Phys. **227**, 97 (2008)
3. F. Radu et al., Nat. Commun. 3, **715** (2012)
4. S. Romer et al., Appl. Phys. Lett. **101**, 222404 (2012)

Abstract

The amorphous structure and magnetic exchange coupling between rare-earth and transition-metal elements give rise for peculiar magnetic order (e.g., spero-, aspero-, and sperimagnetism) in thin amorphous alloy films. The sperimagnetic configuration with its non-collinear ferrimagnetic spin structure occurs in amorphous Fe–Tb alloy films with perpendicular magnetic anisotropy. Within this study, thin amorphous Fe–Tb alloy films were investigated according to their structural and magnetic properties. Beside stoichiometry-dependent changes of the net magnetization and coercivity, strong variations in the reversal process for Fe and Tb dominated alloy films were found mainly in the vicinity of the compensation point. Even for fully compensated magnetic films a magnetization reversal driven by a spin-flop transition is observed. This originates most likely from an orientational moment distribution existing in the Fe and Tb sublattices, which allows an interaction with high external magnetic fields. Furthermore, arrays of percolated Fe–Tb nanodots with pillar diameters of 30 nm and a period of 60 nm became analyzed. Despite the exchange interaction to the trench material, the nanodots reveal a single domain magnetization state and reverse via a more coherent rotation process as deduced from in-field magnetic force microscopy imaging and angular-dependent magneto-optical Kerr effect measurements. Contrary to this, the reversal of the continuous trench material is dominated by domain wall motion and the coercive field becomes enhanced due to pinning effects caused by the nanodot array.

In the second part of this work the exchange interaction and magnetization reversal processes in exchange-biased heterostructures consisting of amorphous ferrimagnetic Fe–Tb alloy films and ferromagnetic Co/Pt multilayers were investigated. The dependence of the interfacial exchange coupling on the stoichiometry and thickness of the Fe–Tb layer as well as the number of repetitions in the Co/Pt multilayers were analyzed. The net magnetization and effective magnetic anisotropy of the Fe–Tb alloy films have an influence on the exchange energy per unit area. A large exchange-bias field up to 8 kOe is found to be accompanied by an interfacial domain wall as probed by element-specific X-ray magnetic circular dichroism absorption measurements. This domain wall exhibits a total thickness between 3 and 4 nm and affects strongly the magnetization reversal in the heterostructure. Additionally, a novel kind of training effect was observed, where the exchange-bias field increases by about 3.5 % from the first to the second field cycle at 10 K. Finally, the interlayer exchange coupling through a Pt spacer layer in Fe–Tb/Pt/[Co/Pt]$_{10}$ heterostructures was probed. High resolution transmission electron microscopy revealed continuous Pt layers with thicknesses of more than 0.8 nm, which provide only indirect exchange coupling between the Fe–Tb film and the Co/Pt multilayer due to spin-polarization of the Pt. Thinner spacer layers cover partially the Co/Pt multilayers and the nucleation of a 3-nm-thick interfacial domain wall provides an exchange-bias field similar to heterostructures without spacer layer.

Acknowledgments

All the successful work concerning this thesis would have never been possible without the help and support of many people. I appreciated the exciting time within the group of surface and interface physics in the last 4 years. The nice atmosphere present in the group during the coffee breaks, birthday celebrations, BBQs, and excursions always provided motivation for the scientific work, which I'm very thankful about.

My gratitude extends in particular to:

- Prof. Manfred Albrecht for his scientific supervision and financial support giving me the opportunity to work on this fascinating research topic as well as for new ideas provided by comprehensive discussions.
- Birgit Hebler for her contribution to this Ph.D. thesis concerning sample preparation and investigation during her Diploma thesis.
- Florin Radu for fruitful discussions and the opportunity of sample preparation at HZB.
- Sri Sai Phani Kanth Arekapudi, Patrick Reinhardt, and Karl Gündel for the contribution to this thesis with regard to their Master, Bachelor, and Diploma thesis, respectively.
- Gunter Beddies and Christoph Brombacher for their guidance during my Diploma thesis and concerning physical as well as general questions.
- Andreas Liebig for XRD measurements and fruitful discussions.
- Denys Makarov for fruitful discussions.
- Beate Mainz, Thomas Wächtler, Herbert Schletter, Marc Lindorf, and the team from the "Strukturlabor" for the TEM investigation.
- Patrick Matthes for measuring XRR and help during several beamtimes at HZB.
- Marcus Daniel for performing RBS measurements.
- Elke Weisse for comprehensive support in the lab.
- Fabian Ganss and Dennis Nissen for their help during several beamtimes at HZB.
- Torsten Kachel and Radu Abrudan for versatile beamtime support.
- Torbjörn Eriksson for providing pre-patterned substrates.
- Thomas Werner for his help and expertise in sputter etching.
- Thomas Mühl for in-field MFM investigations.
- Ute Vales for providing overall administrative support.

- The mechanical and electronic workshop for doing a lot of work especially during the development of the SMBE deposition system.
- Stefan Krause for fruitful discussions and ideas from a different point of view.
- My family and Steffi for their understanding and support during development and research with regard to this Ph.D. thesis in the last 4 years.

Contents

Abbreviations

AF	Antiferromagnetic
AFM	Atomic force microscopy
DCD	Direct current demagnetization
DW	Domain wall
EB	Exchange-bias
ECC	Exchange coupled composite
EELS	Electron energy-loss spectroscopy
F	Ferromagnetic
FI	Ferrimagnetic
FIB	Focused ion beam
FWHM	Fullwidth at half maximum
HAMR	Heat-assisted magnetic recording
HZB	Helmholtz-Zentrum Berlin
IDW	Interfacial domain wall
IRM	Initial remanent magnetization
MBE	Molecular beam epitaxy
MEMS	Microelectromechanical systems
MFM	Magnetic force microscopy
ML	Monolayer
MOKE	Magneto-optical Kerr effect
NIL	Nano imprint lithography
PMA	Perpendicular magnetic anisotropy
PMMA	Polymethylmethacrylat
PPM	Percolated perpendicular media
RBS	Rutherford backscattering spectrometry
RE	Rare-earth
RT	Room temperature
SEM	Scanning electron microscopy
SFD	Switching field distribution
SQUID	Superconducting quantum interference device
TEM	Transmission electron microscopy

TM	Transition-metal
UHV	Ultra high vacuum
XMCD	X-ray magnetic circular dichroism
XRD	X-ray diffraction
XRR	X-ray reflectometry

Symbols

A Integrated dichroic difference intensity of the L_3-edge
A_{eff} Effective exchange stiffness
$A_{\mathrm{Fe,Tb}}$ Atomic numbers of Fe and Tb
a_{hkl} Distance of lattice planes
$\alpha(E)$ Energy-dependent absorption of radiation in a material
B Integrated dichroic difference intensity of the L_2-edge
β Opening angle of an objective
c Speed of light
$\underline{D}_{\mathrm{a,b}}$ Distribution of local anisotropy for rare-earth and transition-metal sites, respectively
$d_{\mathrm{c}\perp}$ Cross-sectional crystallite size
$d_{\mathrm{c}\perp}$ Lateral crystallite size
$d_{\mathrm{c}\perp}$ Cross-sectional crystallite size
δ Exchange length
$\underline{E}$ Electric field
E Photon energy
E_{Ion} Kinetic energy of a backscattered ion
$E_{0,\mathrm{Ion}}$ Incident kinetic energy of an ion
E_{kin} Kinetic energy
$\gamma_{\mathrm{Fe,Tb}}$ Average opening angle of the magnetic moment distribution for Fe and Tb
$\underline{H}$ Magnetic field
$\mathcal{H}$ Hamiltonian
H_{appl} Applied external magnetic field
H_{EB} Exchange-bias field
ΔH_{EB} Change of the exchange-bias field between the first and second field cycle
$H_{\mathrm{S}(\delta)}$ Angle dependent switching field
$H_{\mathrm{S,Co/Pt}}$ Switching field of the Co/Pt multilayer
$H_{\mathrm{S,Fe-Tb}}$ Switching field of the Fe–Tb layer
H_{r} Reversal field
$I(E)$ Intensity of the transmitted beam with photon energy E
$I_0(E)$ Intensity of the incident beam with photon energy E

$j^{\pm}$	Symbol for the spin-orbit split levels with $l+s$ and $l-s$, respectively
$J_{aa',ab,bb'}$	Exchange constant among rare-earth and transition-metal sites
J_{EB}	Interfacial exchange energy density
$\underline{k}$	Incident wave vector
$\underline{k}'$	Scattered wave vector
K_{eff}	Effective magnetic anisotropy
K_S	Magnetic anisotropy induced by the interface per unit area
K_V	Volume anisotropy
l	Azimuthal quantum number
l_c	Azimuthal quantum number of the core level state
$\langle L_z \rangle$	Projected orbital moment along the z-direction
λ	Wave length
λ_S	Saturation magnetostriction constant
$\underline{M}$	Magnetization
m	Mass
m_e	Rest mass of the electron
M_F	Saturation magnetization of the ferromagnetic layer
m_{Ion}	Ion mass
$m_{Nucleus}$	Nucleus mass
M_R	Remanence magnetization
M_S	Saturation magnetization
$\mu(E)$	Energy-specific absorption coefficient of a material
$\mu(E)^{\pm}$	Absorption coefficient for right (+) and left (−) circularly polarized light
N	Integer number
n	Principal quantum number
n'	Refraction index
n_e	Number of occupied states in the valence shell
p_s	Degree of spin polarization
Φ	Deposition rate
$\underline{R}$	Reciprocal lattice vector
ρ	Mass density
S	Area
s	Spin quantum number
$\underline{S}_{a,b}$	Spin of the rare-earth and transition-metal sites, respectively
$\langle S_z \rangle$	Projected spin moment along the z-direction
t	Film thickness
t_F	Thickness of the ferromagnetic layer
t_{Fe-Tb}	Thickness of the Fe–Tb layer
t_{IDW}	Thickness of the interfacial domain wall
$t_{IDW,Co/Pt}$	Thickness of the interfacial domain wall in the Co/Pt multilayer
$t_{IDW,Fe-Tb}$	Thickness of the interfacial domain wall in the Fe–Tb layer
T_N	Néel temperature
$\langle T_z \rangle$	Magntic dipol operator

τ	Time
θ	Incident angle
θ_{hkl}	Glancing angle fulfilling Bragg condition
ϑ	Backscattering angle
V	Volume
ΔV	Volume change
ω	Angle of the sample plane with respect to the detector
$\Delta\omega_{hkl}$	Full width of half maximum of the rocking curve peak
x	Tb content
z	Coordinate perpendicular to the film plane

Chapter 1
Introduction

Collective magnetism can not exist within an amorphous structure as it possesses no long range order. Although Brenner et al. [1] found ferromagnetic properties in amorphous Co–P and Ni–P films already in 1950, this assumption was advocated by the physics community until 1970. However, with the development of more and more thin film deposition techniques [2–6] (e.g., electron beam evaporation, magnetron sputtering, laser ablation, ...) and cryotechniques the controlled fabrication of amorphous magnetic materials [6–8] like transition-metal based amorphous alloy films became capable and the existence of amorphous magnetism accepted.

Beside many different compounds, the class of amorphous rare-earth-transition-metal alloy films reveals manyfold magnetic configurations originating from the amorphous structure and the exchange coupling between the 4f orbitals of the rare-earth element and the 3d orbitals of the transition-metal. In particular, alloy films with heavy rare-earths, whose magnetic moments couple antiparallel to the transition-metal moments [9], possess interesting non-collinear spero-, aspero-, or sperimagnetic configurations [10–12].

Within the first part of this work amorphous sperimagnetic Fe–Tb alloy films with perpendicular magnetic anisotropy were investigated paying special attention to the influence of temperature and external magnetic field on the non-collinear magnetic moments of the Fe and Tb sublattices. For this purpose Fe–Tb alloy films were deposited using magnetron co-sputtering at room temperature. Magnetic and structural characterization provide information about intrinsic magnetic properties and magnetization reversal processes. Particularly, the sublattice reversal mechanism in the vicinity of the compensation point became analyzed by element specific hysteresis loops obtained from high field X-ray magnetic circular dichroism absorption measurements.

Motivated by the peculiar magnetic behavior according to the sperimagnetic configuration of thin continuous amorphous Fe–Tb alloy films, percolated nanodot arrays were investigated with regard to magnetization reversal processes, coupling phenomena, and local magnetic properties. The fabrication of the Fe–Tb nanodot arrays with different stoichiometry was realized by co-deposition of Fe and Tb onto

C. Schubert, *Magnetic Order and Coupling Phenomena*, Springer Theses,
DOI: 10.1007/978-3-319-07106-0_1,

pre-patterned substrates using magnetron sputtering and nano imprint lithography. In this manner, nanodots with a size of 30 nm and a period of 60 nm were generated. Structural and magnetic properties became analyzed by transmission and scanning electron microscopy as well as in-field magnetic force microscopy. Furthermore, angle dependent Kerr magnetometry allowed to obtain information about the intrinsic magnetization reversal processes of the Fe–Tb nanodots and trench material.

Beside common exchange-bias systems [13–16] based on a ferromagnetic film with an antiferromagnetic pinning layer also exchange coupled ferromagnetic/ ferrimagnetic and ferrimagnetic/ferrimagnetic bilayer systems found general interest in the magnetic community within the last decades [17–24]. In particular, heterostructures with perpendicular magnetic anisotropy, in which a ferrimagnetic amorphous rare-earth-transition-metal alloy like Fe–Tb is in the proximity of a transition-metal magnet, exhibit peculiar coupling interactions at the interface [25]. To elucidate this, exchange-biased heterostructures consisting of amorphous ferrimagnetic Fe–Tb alloy films and ferromagnetic Co/Pt multilayers were investigated. The dependence of the interfacial exchange coupling on the stoichiometry and thickness of the Fe–Tb layer as well as the number of repetitions in the Co/Pt multilayers were analyzed. Beside structural investigations and the measurement of integral magnetic properties element specific X-ray magnetic circular dichroism absorption experiments yield information about the magnetic sublattice configurations within different external fields and temperatures. This allows a detailed interpretation of the reversal processes and the dependency of the exchange-bias field present in the exchange coupled heterostructures.

Finally, the influence of the interface morphology and coupling constant on the exchange-bias field was explored for Fe–Tb/Pt/$[Co/Pt]_{10}$ heterostructures with different Pt spacer layer thicknesses. A comprehensive high resolution transmission electron microscopy study was realized providing valuable information for the interpretation and modeling of the magnetic data.

References

1. A. Brenner, D.E. Couch, E.K. Williams, J. Res. Nat. Bur. Stand. **44**, 109 (1950)
2. P. Kelly, R. Arnell, Vacuum **56**, 159 (2000)
3. K. Wasa, *Handbook of Sputter Deposition Technology* (Elsevier, Amsterdam, 2012). ISBN 978-1-437-73484-3
4. M.D. Shirk, P.A. Molian, J. Laser Appl. **10**, 18 (1998)
5. C. Phipps, *Laser Ablation and its Applications, Springer Series in Optical Sciences* (Springer, Boston, 2007). ISBN 978-0-387-30452-6
6. J.M.D. Coey, *Magnetism and Magnetic Materials* (Cambridge University Press, New York, 2010). ISBN 978-0521816144
7. K.H.J. Buschow, *Handbook of Magnetic Materials*, vol. 6 (Elsevier, Amsterdam, 1991), ISBN 978-0-444-88952-2
8. G. Connell, D.S. Bloomberg, *Amorphous Rare-Earth Transition-Metal Alloys* (Springer, US, 1985). ISBN 978-1-4612-9519-8
9. I. Campbell, J. Phys. F Met. Phys. **2**, L47 (1972)

10. J. Coey, J. Chappert, J. Rebouillat, T. Wang, Phys. Rev. Lett. **36**, 1061 (1976)
11. A. Andreenko, S. Nikitin, Phys. Usp. **40**, 581 (1997)
12. K. Handrich, S. Kobe, *Amorphe Ferro- und Ferrimagnetika* (Akademie-Verlag Berlin, 1980), ISBN 978-3-87664-044-0
13. J. Nogués, I. Schuller, J. Magn. Magn. Mater. **192**, 203 (1999)
14. M. Kiwi, J. Magn. Magn. Mater. **234**, 584 (2001)
15. A. Berkowitz, K. Takano, J. Magn. Magn. Mater. **200**, 552 (1999)
16. R. Stamps, J. Phys. D Appl. Phys. **33**, R247 (2000)
17. W.C. Cain, M. Kryder, J. Appl. Phys. **67**, 5722 (1990)
18. N. Smith, W.C. Cain, J. Appl. Phys. **69**, 2471 (1991)
19. P.P. Freitas, J.L. Leal, L.V. Melo, N.J. Oliveira, L. Rodrigues, A.T. Sousa, Appl. Phys. Lett. **65**, 493 (1994)
20. R. Sbiaa, H. Le Gall, J. Appl. Phys. Lett. **75**, 256 (1999)
21. F. Canet, S. Mangin, C. Bellouard, M. Piecuch, Europhys. Lett. **52**, 594 (2000)
22. F. Canet, S. Mangin, C. Bellouard, M. Piecuch, A. Schuhl, J. Appl. Phys. **89**, 6916 (2001)
23. S. Mangin, F. Montaigne, A. Schuhl, Phys. Rev. B **68**, 140404 (2003)
24. F. Radu, R. Abrudan, I. Radu, D. Schmitz, H. Zabel, Nat. Commun. **3**, 715 (2012)
25. S. Mangin, T. Hauet, P. Fischer, D. Kim, J.B. Kortright, K. Chesnel, E. Arenholz, E.E. Fullerton, Phys. Rev. B **78**, 024424 (2008)

Chapter 2
Thin Amorphous Fe–Tb Alloy Films

Until the mid 20th century the spontaneous magnetization was believed to require long range order, which is only given in crystalline materials. Interestingly, in 1946 H. König [1] found ferromagnetism in thin Fe films, which were deposited on cooled substrates. Although electron diffraction measurements revealed an amorphous structure the ferromagnetic behavior was traced to small crystallites embedded in the film. Since 1970 the existence of amorphous magnetic materials is proofed and commonly accepted by the physical community. By today comprehensive knowledge about the origin of amorphous magnetism exists. Many different systems are discussed in the literature [2–5]. On this account the present chapter focuses on the discussion of thin amorphous Fe–Tb alloy films as particular system important concerning this work.

2.1 Phase Diagram

The binary rare-earth-transition-metal alloy system consisting of terbium and iron exhibits four stable crystalline phases $Fe_{17}Tb_2$, $Fe_{23}Tb_6$, Fe_3Tb, and Fe_2Tb as outlined in the phase diagram in Fig. 2.1 for bulk samples in thermal equilibrium [6–8]. The ferromagnetic (F) $Fe_{17}Tb_2$ and $Fe_{23}Tb_6$ phases coexist in the composition range from 10.5 to 20.7 at.% Tb together with α-Fe. For a Tb content higher than 20.7 at.% and smaller than 25 at.% mixed phases arise consisting of α-Fe and ferrimagnetic (FI) Fe_3Tb clusters. Higher portions of Tb produce mixed phases of α-Fe and FI Fe_2Tb up to compositions around 33 at.% Tb. Outside the described composition range a mixture of α-Fe and α-Tb clusters exist. The crystal structure, Curie temperature, and magnetic configuration of all four crystalline phases are listed in Table 2.1 [9].

The formation of stable crystalline phases in the composition range from 10.5 at.% to around 33 at.% Tb occurs only for bulk material coming from the liquid phase. Thin films behave different. Using magnetron co-sputtering, a common technique to produce thin metallic alloy films, deposited Fe–Tb layers reveal an amorphous structure [10–12]. At room temperature (RT) the amorphous structure is stable, since the

C. Schubert, *Magnetic Order and Coupling Phenomena*, Springer Theses,
DOI: 10.1007/978-3-319-07106-0_2,

Table 2.1 Structural and magnetic properties of crystalline phases in the Fe–Tb alloy system [9]

Compound	Symmetry	Structure	Curie temperature (K)	Magnetic structure
$Fe_{17}Tb_2$	Hexagonal	$Ni_{17}Th_2$	408	F
$Fe_{23}Tb_6$	Cubic	$Mn_{23}Th_6$	574	F
Fe_3Tb	Rhombohedral	Ni_3Pu	648–655	FI
Fe_2Tb	Cubic	Cu_2Mg	696–711	FI

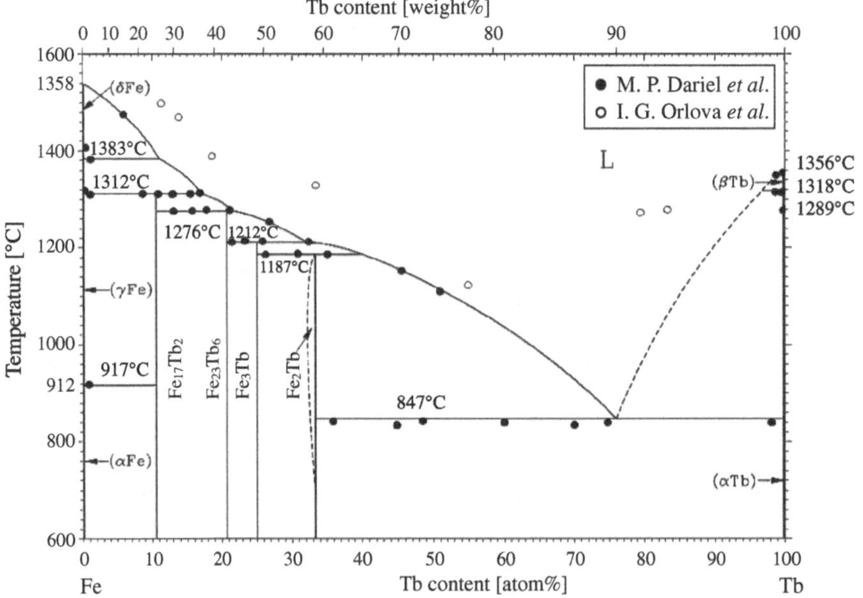

Fig. 2.1 Binary phase diagram of the $Fe_{100-x}Tb_x$ alloy system based on data from T. B. Massalski et al. [6] supplemented by several data points from M. P. Dariel et al. [7] and I. G. Orlova et al. [8] obtained from bulk samples in thermal equilibrium

crystallization in this system takes place not below 200 °C as it was found by N. Sato [13] for 1 μm thick Fe–Tb films. An overview about the composition dependence of the crystallization temperature is shown in Fig. 2.2. The temperature required for crystallization increases monotonous towards higher Tb contents. Although the literature provides no detailed discussion onto this, the enhancement of the crystallization temperature can be most likely attributed to the different phase mixtures present in this composition region.

Concerning the purpose of this work, the thin amorphous Fe–Tb alloy films were not exposed to temperatures higher than 450 K. Therefore, crystallization will be excluded in following annotations and discussions.

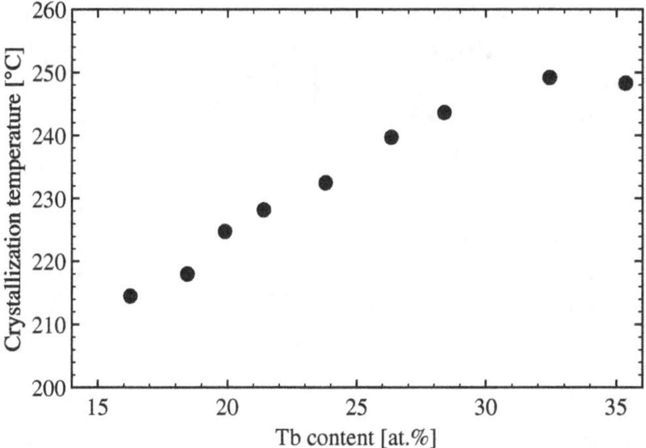

Fig. 2.2 Crystallization temperature of 1-μm-thick amorphous Fe–Tb alloy films as a function of the composition [13]

2.2 Intrinsic Magnetic Structure: Magnetic Anisotropy, Exchange Coupling, and Sperimagnetism

The common structures in magnetic materials, whether F, FI or antiferromagnetic (AF), are related to long range order and thus to crystalline structures. Although amorphous materials possess structural disorder collective magnetism exists. The chemical and structural disorder in this materials lead to a distribution in magnetic moment and exchange interaction. Furthermore, the varying interatomic distances induce random electrostatic fields giving rise to locally varying single ion anisotropy. In this manner the magnetic structure arises from the competition between exchange interaction trying to align magnetic moments and the local anisotropies due to the local structural order [3, 4, 14–16]. This can be described by the following model Hamiltonian for two subnetwork systems in particular valid for rare-earth-transition-metal alloys: [15, 17]

$$\mathcal{H} = -\sum \underline{D}_a S_{a,z}^2 - \sum \underline{D}_b S_{b,z}^2 - \sum J_{aa'} \underline{S}_a \cdot \underline{S}_{a'} - \sum J_{ab} \underline{S}_a \cdot \underline{S}_b - \sum J_{bb'} \underline{S}_b \cdot \underline{S}_{b'}, \quad (2.1)$$

where a and b refer to different sites on the rare-earth and transition-metal subnetworks, respectively, $\underline{D}_{a,b}$ denotes the distribution of local anisotropy, $\underline{S}_{a,b}$ the spin, and $J_{aa',...}$ the exchange constant. Please note, z as index denotes the z-component of the spin. Normally the contributions $\underline{D}_b$ and $J_{aa'}$ are negligible. Depending on the values for the exchange and the local anisotropy asperomagnetic, speromagnetic and sperimagnetic configurations are possibles as outlined in Fig. 2.3.

In case of the Fe–Tb alloy system a sperimagnetic configuration arises. The magnetic exchange between the Fe and Tb moments occurs via $3d-5d-4f$ hybridization

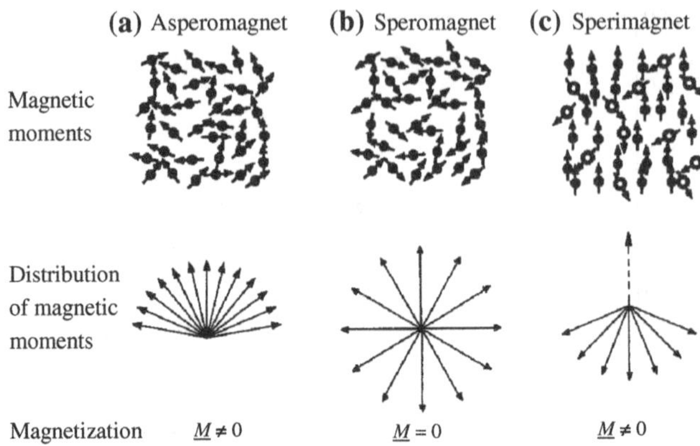

Fig. 2.3 Different magnetic configurations in amorphous one and two component systems. **a** Asperomagnet possessing a non-zero magnetization with a moment distribution pointing in a preferential direction. **b** Speromagnet exhibiting a zero magnetization due to a isotropic moment distribution. **c** Two subnetwork system with antiparallel moment distribution named sperimagnet [15]

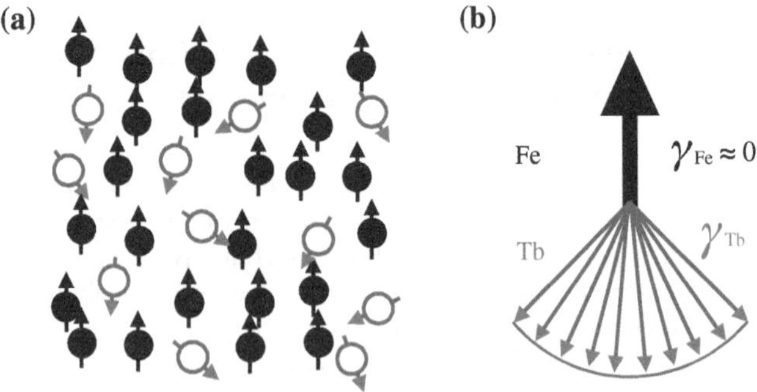

Fig. 2.4 **a** Moment distribution in a sperimagnetic material with antiparallel oriented magnetic sublattices. **b** Average magnetic moment distribution with angle γ in the amorphous Fe–Tb alloy system [15]

leading to an antiparallel alignment $J_{\mathrm{FeTb}} < 0$ as known for heavy rare-earth-transition-metal alloys following Hund's rules [18]. Due to a strong interatomic Fe coupling $J_{\mathrm{FeFe'}} > 0$ and a vanishing local anisotropy $\underline{D}_{\mathrm{Fe}}$ for the iron sites a more or less collinear orientation of the Fe moments is observed. In contrast to this the Tb moments exhibit a weak exchange coupling $J_{\mathrm{TbTb'}} << J_{\mathrm{FeFe'}}$ and a strong distribution of the local anisotropy $\underline{D}_{\mathrm{Tb}}$ producing a broad orientational distribution of the Tb moments as illustrated in Fig. 2.4. This moment distribution is commonly designated as fanning cone.

In thin amorphous Fe–Tb alloy films the growth conditions imprint a perpendicular magnetic anisotropy (PMA) meaning the resulting component of the fanning cone with respect to the Fe sublattice points perpendicular to the substrate plane.

The growth process induces an anisotropy in the short range order as observed by different groups [19–22]. This leads to different next neighbor distances and coordination numbers for Fe–Fe, Fe–Tb, and Tb–Tb pairs in out-of-plane and in-plane direction. In particular, a preference for unlike (i.e. Fe–Tb) near-neighbor pairs to align in out-of-plane direction was found. Due to this chemical anisotropy, the interaction of the local crystal field with the elongated shape of the 4f orbitals via spin-orbit coupling produces a magnetic anisotropy with an easy axis pointing along the out-of-plane direction. Often this is also described as single ion anisotropy of the Tb sites with preferential out-of-plane direction due to the exchange interaction among them and the Fe sublattice. Beside this origin of PMA magnetostriction [23] is known and will be discussed later.

2.3 Magnetic Properties with Regard to Alloy Composition, Film Thickness, and Temperature

Due to its sperimagnetic nature amorphous Fe–Tb alloy films possess manifold magnetic properties with particular temperature and composition dependencies. The present chapter is dedicated to give an overview of the different dependencies mainly focusing on saturation magnetization, coercivity, and magnetic anisotropy.

As previously mentioned heavy rare-earth-transition-metal alloys possess an antiparallel moment structure due to the parallel coupling of the orbital and spin moment in the rare-earth leading to an overall ferrimagnetic behavior. In consequence of the high magnetic moment of rare-earth metal ions (e.g., 9 μ_B per Tb ion) Fe–Tb alloy films are very sensitive to the amount of Tb they contain. Figure 2.5 shows the net saturation magnetization at RT, Curie temperature, and compensation temperature in (a) and the RT coercivity in (b) as a function of the Tb content of the alloy [11]. With increasing amount of Tb in the alloy the net saturation magnetization, which is dominated by the Fe sublattice decreases to zero towards the compensation composition at which the magnetic sublattices of Fe and Tb compensate each other. Above the compensation point the magnetization increases again towards higher Tb contents, while the dominant magnetic sublattice has changed from Fe to Tb. At room temperature the compensation composition is around 23 at.% [11]. Please note, the RT compensation composition can vary with different preparation techniques, since the compensation of the sublattice magnetic moments is strongly related to the intrinsic structure of the amorphous sperimagnetic alloy film. In particular the anisotropy in the chemical short range order and the shape of the fanning cone play an important role as discussed in the previous chapter.

Beside the compensation point at RT for 23 at.% Tb, other compensation temperatures (T_{Comp}) in the range up to the Curie temperature (T_C) of about 400 K are possible. Between 20 and 24 at.% Tb a compensation point exists (see Fig. 2.5a). For lower and higher amounts of Tb the Fe–Tb alloy films reveal a net magnetization

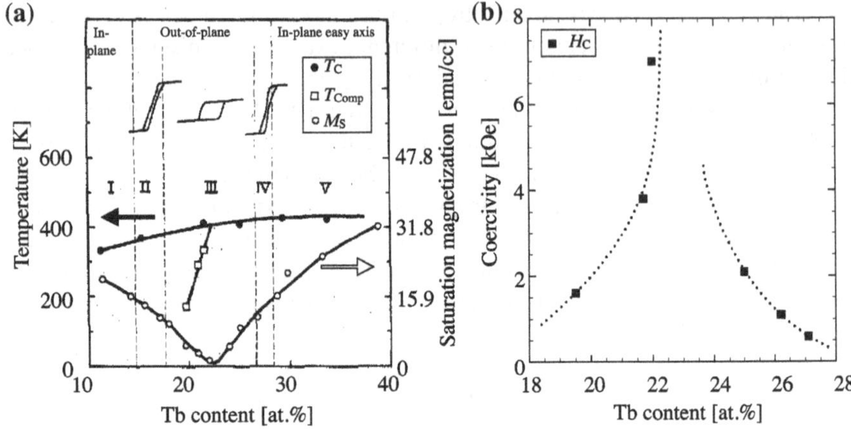

Fig. 2.5 **a** Curie temperature, compensation temperature, saturation magnetization, and **b** coercivity of amorphous Fe–Tb films with varying Tb concentration and a thickness of about 150 nm [11]. Please note that the magnetization and coercivity *curves* were obtained at room temperature. The *solid* and *dashed lines* serve as a guide to the eye. Region I and V correspond to the composition of Fe–Tb films with in-plane magnetic anisotropy. For compositions dedicated to region II, III, and IV the alloy films exhibit out-of-plane anisotropy, while the partition into the three regions takes the different shaped hysteresis loops into account. The increasing coercivity in (**b**) is caused by the RT compensation composition of about 23 at.%

dominated solely by the Fe and Tb sublattice, respectively, over the whole temperature range and no compensation point exists.

Concerning the Curie temperature only a slight dependency with regard to the alloy composition was observed. It decreases from about 400 to 350 K towards smaller amounts of Tb as presented in Fig. 2.5a. In comparison to pure crystalline Fe with $T_C \approx 1043$ K the values for amorphous Fe–Tb alloy films are significant smaller accounting for the weak exchange coupling between the Fe and Tb moments due to the 3d − 5d − 4f hybridization. The reason for the decrease in T_C towards smaller Tb contents lies in the reduction of the Fe–Fe next neighbor distances due to the less amount of larger Tb atoms in the alloy and the negative contribution of the Fe–Fe coupling term to T_C, as discussed in the literature [12]. Below 10 at.% Tb more and more crystalline bcc Fe grains exist leading to a strong increase in T_C towards the pure Fe value, when the Tb concentration becomes zero (not shown). A similar composition dependency as for the Curie temperature is also observed for the exchange stiffness A in the Fe–Tb alloy films as presented in Fig. 2.6a, since both values are related to each other.

Contrary to the net saturation magnetization the coercivity increases progressively in the vicinity of the compensation composition due to the inverse relation to the magnetization:

$$H_C \approx \frac{K_{\text{eff}}}{M_S}, \tag{2.2}$$

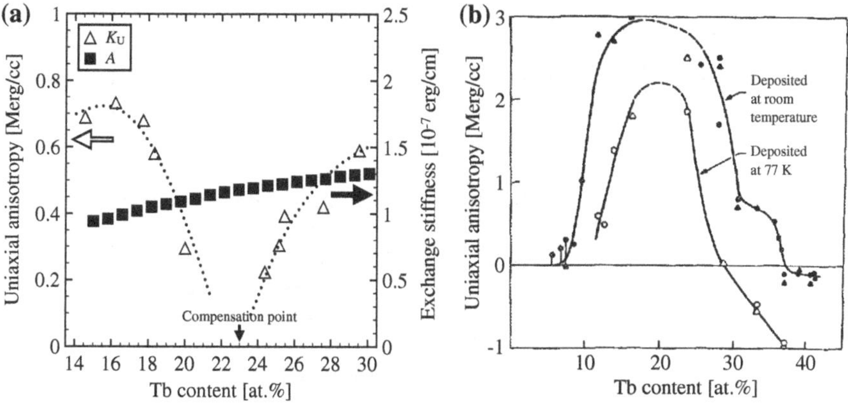

Fig. 2.6 **a** Uniaxial magnetic anisotropy K_U and calculated exchange stiffness A of 0.7 to 1.2 μm thick amorphous Fe–Tb alloy films at 298 K drawn as a function of the Tb content [24]. Details about the calculated data points are given in the text. **b** Uniaxial anisotropy of 500-nm-thick Fe–Tb films depending on the Tb content [25]. *Solid symbols* represent films deposited at room temperature, while open symbols denote films deposited on a liquid-nitrogen-cooled substrate. The *dashed* and *solid lines* serve as a guide to the eye

where H_C, K_{eff}, and M_S denote the coercivity, effective magnetic anisotropy, and net saturation magnetization, respectively. With regard to the anisotropy the given relation provides a useful lower estimation, which is exact within a factor of about 1 for amorphous ferrimagnetic rare-earth-transition-metal alloy films [3].

Amorphous $Fe_{100-x}Tb_x$ alloy films in a composition range of $15\,at.\% \leq x \leq 29\,at.\%$ possess PMA indicated with region II to IV in Fig. 2.5a, while the reversal mechanism changes slightly in the different regions due to the varying net saturation magnetization. Beside the observed out-of-plane easy axis the total value of the anisotropy constant is controversially discussed. Especially in the vicinity of the compensation point the exact behavior is unknown. Minimum as well as maximum values for PMA at the compensation point are reported by several groups as outlined in Fig. 2.6a, b, respectively [24, 25]. This uncertainty is attributed to the manifold peculiar structural properties present in the amorphous alloy related to the different preparation techniques, film thicknesses, and seed layers, which determine the intrinsic magnetic properties like the distribution of the local anisotropy becoming very influential at the compensation point.

The alloy composition is not exclusively influencing the magnetic properties, film thickness plays also an important role. In Fig. 2.7 the coercivity and net saturation magnetization is shown as a function of the thickness of amorphous $Fe_{80}Tb_{20}$ films [13]. With thinner alloy films the saturation magnetization increases while the coercivity decreases, which is related the growth conditions within the first 10-nm-regime where the substantial influence of the substrate produces a strong magnetic in-plane component in the alloy. Thicker layers exhibit a dominant PMA imprinted by the

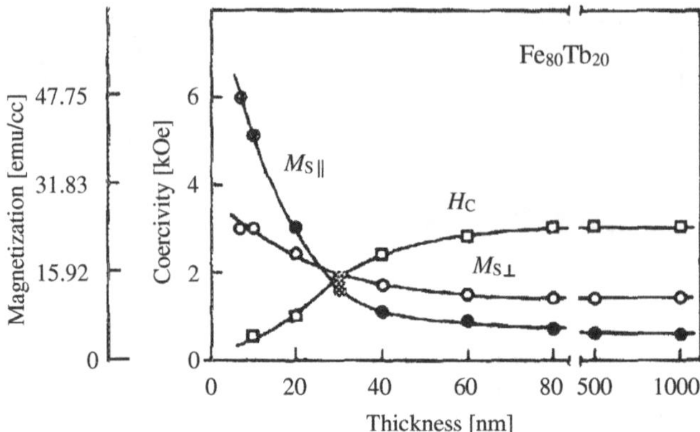

Fig. 2.7 Coercivity and saturation magnetization of $Fe_{80}Tb_{20}$ films at RT with varying thickness [13]. Two different saturation values for the magnetization are given for in-plane and out-of-plane field geometry

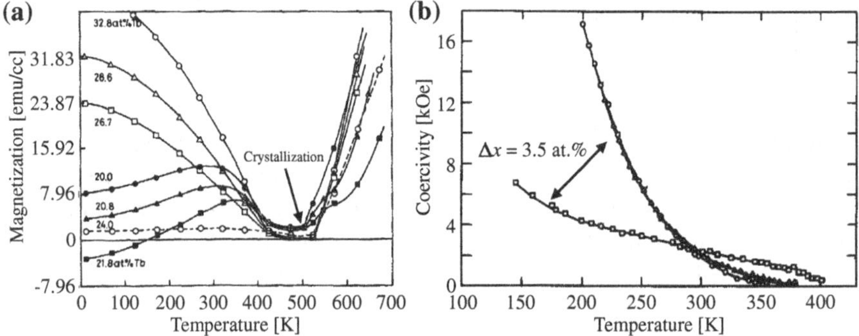

Fig. 2.8 **a** Temperature dependence of the resulting net magnetization of 1-μm-thick Fe–Tb films with different compositions [13]. **b** Coercivity of 100-nm-thick Tb rich Fe–Tb films as a function of temperature [26]. A variation in Tb content of only 3.5 at.% leads to strong differences in the temperature dependency

anisotropy in the chemical short range order due to the directed growth process as mentioned in the previous section.

Furthermore, the sperimagnetic Fe–Tb alloy films show a strong temperature dependence in the net saturation magnetization and coercivity due to the orientational distribution of the Tb moments (see Fig. 2.4b) [13, 26]. In general a strong increase of the Tb sublattice moment towards lower temperatures causing a relative decrease in the net magnetization dominated by the Fe sublattice is observed as outlined in Fig. 2.8a. Depending on the Tb concentration different curve shapes are possible leading to the existence of a compensation point and dominant Tb sublattice magnetization in the lower temperature regime. Figure 2.8b presents the temperature dependence of the coercivity of 100-nm-thick Tb dominated Fe–Tb films.

The increase with decreasing temperature originates from the strong enhancement of magnetic anisotropy, which also strongly depends on the alloy stoichiometry.

Following the known properties of the amorphous Fe–Tb alloy system with regard to the frame of this work $Fe_{100-x}Tb_x$ films with thicknesses between 5 nm and 50 nm in a composition range from 14 to 30 at.% Tb were produced using magnetron co-sputtering technique.

2.4 Magnetostriction

The magnetostriction as intrinsic behavior of magnetic materials describes the influence of the magnetization onto the structural properties and *vice versa*. Generally, the structure interacts via crystal field and spin-orbit coupling with the magnetization. Thus, changes in the direction of the intrinsic magnetic moments may produce changes in the inter atomic distances inducing stress and deformations in the material. The volume or length change between the saturated and demagnetized state is denoted as saturation magnetostriction constant λ_S and can be deduced from the size difference normalized to the initial size in demagnetized state $\lambda_S = \frac{\Delta V}{V}$.

The highest value of magnetostriction, of the order of 2.5×10^{-3} was found in the crystalline Laves phase of $TbFe_2$ by A. E. Clark and H. S. Belson in 1972 [27]. Compared to elemental Co with a magnetostriction constant of about 0.06×10^{-3} this Laves phases possess a very strong influence of the magnetic field on the structure. However, this alloys exhibit also strong magnetocrystalline anisotropy making them unsuited for magnetoelastic applications due to the high saturation fields they require. Therefore, extensive research was done concerning amorphous rare-earth-transition-metal alloys [9]. Since the magnetostrictive effects in this amorphous materials are related to magnetoelastic interactions between local strains and anisotropies controlling the direction of magnetic moment, relatively high magnetostriction constants with moderate magnetic anisotropies are possible already at RT [4]. Particularly, amorphous Fe–Tb alloy films exhibit magnetostriction constants between 0.1×10^{-4} and 1×10^{-4} varying with the Tb concentration as outlined in Fig. 2.9 [23]. The high magnetostrictive values take place above 30 at.%, which goes along with the strain induced transition of the easy axis from out-of-plane to in-plane direction. This adjustability of anisotropy and magnetostriction with the Tb content makes this system suitable for application.

Concerning the purpose of this work magnetostriction plays a minor role, since all films were prepared with thicknesses less than 50 nm on thick non-elastic substrates keeping the magnetostrictive influence rather small. Furthermore, the used $Fe_{100-x}Tb_x$ alloy films in a composition range from 14 to 30 at.% Tb reveal changes in the magnetostriction constant less than half an order of magnitude (see Fig. 2.9).

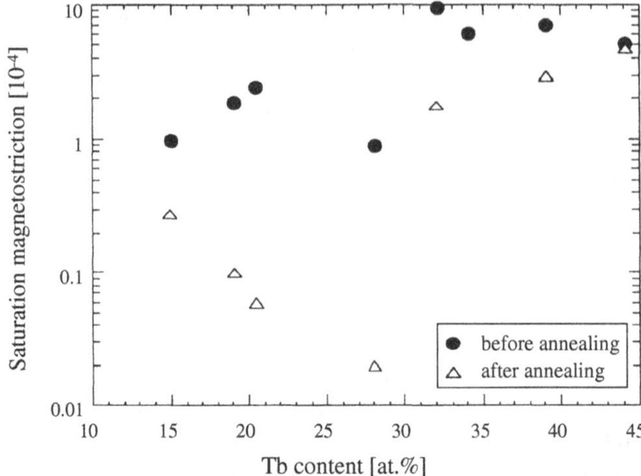

Fig. 2.9 Saturation value of the magnetostriction coefficient of amorphous $Fe_{100-x}Tb_x$ alloy films depending on the Tb content for as grown and annealed films [23]

2.5 Fe–Tb Films for Application

In the past thin amorphous Fe–Tb alloy films were used in magneto-optical data storage devices due to its high magnetic anisotropy and the ability to form stable bubble domains [28–30]. The relatively low Curie temperature at around 400 K of the alloy system provide the ability for technical suitable Curie point writing where a short laser pulse heats the magnetic film locally to T_C giving rise to bubble domain formation induced by the stray field and the guided external applied magnetic field when the temperature becomes reduced below T_C again [11, 30]. As one data bit is represented by one bubble domain the storage density is limited to the stability condition and minimum radius of the bubble with respect to the material [31, 32]. Additionally, the optical resolution of the laser spot is limited by its wave length. Both limitations restrict the achievable maximum areal density. For this reason magneto-optical data storage was overtaken by magnetic data storage in granular media using conventional inductive write heads as superior in performance. Due to this magnetic data storage is now the common storage technic available on the market.

Since nowadays the areal density in CoCrPt based perpendicular granular media approaches more and more to its maximum limit (1 Tbit/inch^2) due to the super-paramagnetic effect [33–35], new storage concepts come up providing new routes to achieve ultra high storage densities beyond 1 Tbit/inch^2. One of these concepts is heat assisted magnetic recording (HAMR) where a localized laser heat pulse provides writability (near the Curie temperature) for high anisotropic magnetic materials like ordered $L1_0$ FePt [36] used in order to guarantee magnetization stability in grains with sizes below 5 nm [37, 38].

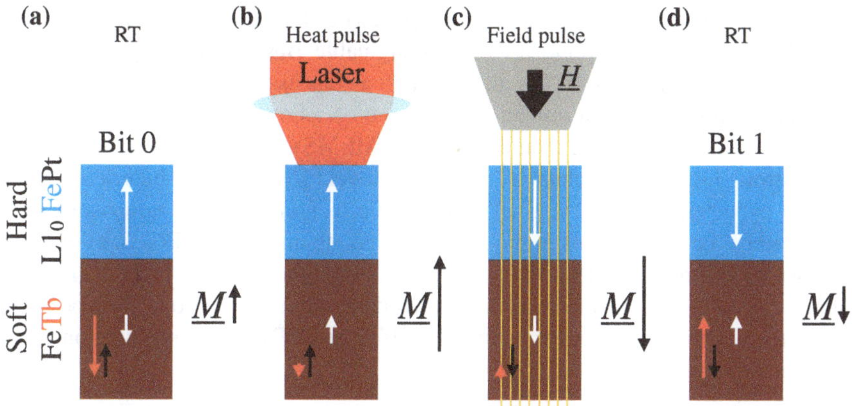

Fig. 2.10 Schematic drawing of the heat assisted reversal process in ECC media with soft Fe–Tb as functional layer. **a** The Fe–Tb layer is dominated by the Tb sublattice moment at RT. **b** The laser pulse heats the Fe–Tb layer above the compensation temperature and the net magnetization becomes dominated by the Fe sublattice leading to a parallel alignment to the hard $L1_0$ FePt layer. **c** The ECC layer stack with enhanced saturation magnetization becomes reversed by a magnetic field pulse via domain nucleation in the soft Fe–Tb layer. **d** After cooling down to RT the reversed magnetization state in the FePt layer becomes stabilized by the Tb dominated antiparallel oriented Fe–Tb layer. After this cycle the data bit is changed from zero to one

With regard to HAMR the temperature dependence of the magnetization is a key property needed to be controlled and adjusted to the purposes of the storage application. Thus, amorphous Fe–Tb alloy films represent a promising candidate for the use as a functional layer for HAMR media as the magnetization can be controlled easily by the Tb concentration and ambient temperature (see Sect. 2.3). In doing so, Fe–Tb alloy films can be utilized as soft magnetic layer in connection to a hard magnetic storage layer like $L1_0$ FePt forming an exchange coupled composite (ECC) media. This ECC media as suggested by Suess et al. [39, 40] provides high temperature stability due to the anisotropy of the hard magnet and overcomes the writability issue by nucleation of a domain wall in the soft magnet during the reversal as a sufficiently high switching field is applied. Concerning the Fe–Tb layer its FI structure stabilizes the magnetic configuration further by interfacial exchange coupling. In particular the temperature dependence of the magnetization can be adjusted so that the compensation point lies above RT. On one hand, this allows an antiparallel alignment between the Tb dominated Fe–Tb layer and the hard magnet at RT lowering the contribution of the stray field, which provides the mentioned bit stabilization. And on the other hand, at the high temperature state after heat pulse the net moment of the Fe–Tb layer becomes Fe dominant and aligns parallel to the moment of the hard magnet providing high saturation magnetization for the switching of the ECC layer stack. A schematic drawing for the heat assisted reversal of the proposed ECC media with functional Fe–Tb soft magnetic layer is presented in Fig. 2.10a–d. Please note, the readback of such ECC media is still an issue and requires innovative read concepts.

Beside the concept about the use of amorphous Fe–Tb alloy films in ultra high density HAMR media further applications are imaginable. Based on the magnetostriction present in the amorphous Fe–Tb alloy system as discussed in the previous chapter one can consider this material for micro force or vibration sensors following the approach of microelectromechanical systems (MEMS) with magnetic readout. Also a very strong interfacial exchange coupling in novel FI/F exchange-bias heterostructures is expected for this class of material due to non-existing grain boundaries, atomic steps, or dislocations at the interface making amorphous Fe–Tb alloy films feasible for magnetic field sensor and spin valve applications based on the giant and tunnel magneto resistance effect. A detailed discussion on this topic will be given within the chapter about the exchange-bias effect in FI/F rare-earth-transition-metal based heterostructures.

References

1. H. König, Naturwissenschaften **33**, 71 (1946)
2. S. Kobe, A.R. Ferchmin, J. Mater. Sci. **12**, 1713 (1977)
3. K. Handrich, S. Kobe, *Amorphe Ferro- und Ferrimagnetika* (Akademie, Berlin, 1980), ISBN 978-3-87664-044-0
4. K.H. Buschow, *Handbook of Magnetic Materials*, vol. 6 (Elsevier, Amsterdam, 1991), ISBN 978-0-444-88952-2
5. J.A. Fernandez-Baca, *The magnetism of amorphous metals and alloys* (World Scientific, Singapore, 1995), ISBN 978-9-810-21033-5
6. T.B. Massalski, H. Okamoto, P.R. Subramanian, L. Kacprzak, *Binary Alloy Phase Diagrams* (ASM International, Materials Park, 1990). ISBN 978-0-87170-403-0
7. M.P. Dariel, J.T. Holthuis, M.R. Pickus, J. Common Met. **45**, 91 (1976)
8. I.G. Orlova, A.A. Eliseev, G.E. Chuprikov, E. Rukk, Zhurnal Neorganicheskoi Khimii **22**, 2557 (1977)
9. E. Wohlfarth, *Handbook of Magnetic Materials*, vol 1 (North-Holland Publishing Company, Amsterdam, 1980), ISBN 978-0-444-85311-0
10. J. Rhyne, J. Schelleng, N. Koon, Phys. Rev. B **10**, 4672 (1974)
11. Y. Mimura, N. Imamura, T. Kobayashi, IEEE Trans. Magn. **12**, 779 (1976)
12. Y. Mimura, N. Imamura, Appl. Phys. Lett. **28**, 746 (1976)
13. N. Sato, J. Appl. Phys. **59**, 2514 (1986)
14. J. Coey, J. Chappert, J. Rebouillat, T. Wang, Phys. Rev. Lett. **36**, 1061 (1976)
15. J. Coey, J. Appl. Phys. **49**, 1646 (1978)
16. J. Rebouillat, A. Lienard, J. Coey, R. Arrese-Boggiano, J. Chappert, Physica B **86**, 773 (1977)
17. J.J. Rhyne, *Handbook on the Physics and Chemistry of Rare Earths*, vol. 2 (North-Holland Publishing Company, Amsterdam, 1979), ISBN 978-0-444-85021-X
18. I. Campbell, J. Phys. F: Met. Phys. **2**, L47 (1972)
19. C. Robinson, M. Samant, Appl. Phys. A Mater. Sci. Process. (1989)
20. M. Tewes, J. Zweck, H. Hoffmann, J. Magn. Magn. Mater. **95**, 43 (1991)
21. V.G. HarriS, K.D. Aylesworth, B.N. Das, W.T. Elam, N.C. Koon, Phys. Rev. Lett. **69**, 1939 (1992)
22. T. Hufnagel, S. Brennan, P. Zschack, B. Clemens, Phys. Rev. B **53**, 12024 (1996)
23. J. Huang, C. Prados, J. Evetts, A. Hernando, Phys. Rev. B **51**, 297 (1995)
24. Y. Mimura, N. Imamura, T. Kobayash, A. Okada, Y. Kushiro, J. Appl. Phys. **49**, 1208 (1978)
25. R. Van Dover, M. Hong, E. Gyorgy, J. Dillon, S. Albiston, J. Appl. Phys. **57**, 3897 (1985)
26. R. Malmhäll, T. Chen, Thin Solid Films **125**, 257 (1985)

27. A. Clark, H. Belson, Phys. Rev. B **5**, 3642 (1972)
28. M. Kryder, Ann. Rev. Mater. Sci. **23**, 411 (1993)
29. Y. Aoki, IEEE Trans. Magn. **MAG-20**, 1022 (1984)
30. S. Tsunashima, J. Phys. D: Appl. Phys. **34**, R87 (2001)
31. J. Nielsen, Metall. Mater. Trans. B **2**, 625 (1971)
32. J. Nielsen, Ann. Rev. Mater. Sci. **9**, 87 (1979)
33. H.J. Richter, A. Lyberatos, U. Nowak, R.F.L. Evans, R.W. Chantrell, J. Appl. Phys. **111**, 033909 (2012)
34. R. Evans, R.W. Chantrell, U. Nowak, A. Lyberatos, H.J. Richter, Appl. Phys. Lett. **100**, 102402 (2012)
35. S. Piramanayagam, K. Srinivasan, J. Magn. Magn. Mater. **321**, 485 (2009)
36. S. Sun, C.B. Murray, D. Weller, L. Folks, Science **287**, 1989 (2000)
37. M. Kryder, E.C. Gage, T.W. McDaniel, W.A. Challener, R.E. Rottmayer, G. Ju, Y.T. Hsia, M.F. Erden, Proc. IEEE **96**, 1810 (2008)
38. F. Akagi, M. Mukoh, M. Mochizuki, J. Ushiyama, T. Matsumoto, H. Miyamoto, J. Magn. Magn. Mater. **324**, 309 (2012)
39. D. Suess, T. Schref, S. Fähler, M. Kirschner, Appl. Phys. Lett. **87**(1), 012504 (2005)
40. D. Suess, T. Schrefl, R. Dittrich, M. Kirschner, J. Magn. Magn. Mater. **290–291**, 551 (2005b)

Chapter 3
Co/Pt Multilayers

Since the discovery [1] of PMA in artificial layered structures consisting of Co and Pt or Pd intensive research on multilayer structures led to comprehensive knowledge about the magnetic properties and physical origin of PMA with respect to crystal structure and morphology [1–10]. Beside the application point of view were Co/Pt multilayers are promising candidates for future storage materials [11–15] they represent a suitable model system for fundamental investigations on the interfacial exchange coupling and reversal mechanism in magnetic exchange coupled heterostructures with PMA as the present work focuses on. In the following an overview on the magnetic properties of Co/Pt multilayers will be given with a particular discussion about the origin of magnetic anisotropy.

3.1 Origin of Magnetic Anisotropy

Artificial layered structures like Co/Pd or Co/Pt multilayers possess magnetic anisotropy with an easy axis of magnetization pointing perpendicular to the layer planes [1]. A phenomenological description of the effective magnetic anisotropy was given by Draaisma et al. [10] outlined in the following equation:

$$K_{\text{eff}} = 2\frac{K_{\text{S}}}{t_{\text{Co}}} + K_{\text{V}}, \tag{3.1}$$

where K_{eff}, K_{S}, and K_{V} denote the effective magnetic anisotropy, the anisotropy induced by the interface per unit area, and the volume anisotropy, respectively. The volume anisotropy arises from the magnetostatic energy, crystalline anisotropy, and magnetoelastic energy. The thickness of the Co layer is designated as t_{Co}.

The intrinsic physical origin of the interfacial anisotropy perpendicular to the plane was found by Nakahima et al. [3] in 1998. By using X-ray magnetic circular dichroism measurements at the Co $L_{2,3}$ and $M_{4,5}$ core edges a strongly enhanced perpendicular Co orbital moment was observed for Co/Pt multilayers with high PMA.

C. Schubert, *Magnetic Order and Coupling Phenomena*, Springer Theses,
DOI: 10.1007/978-3-319-07106-0_3, © Springer International Publishing Switzerland 2014

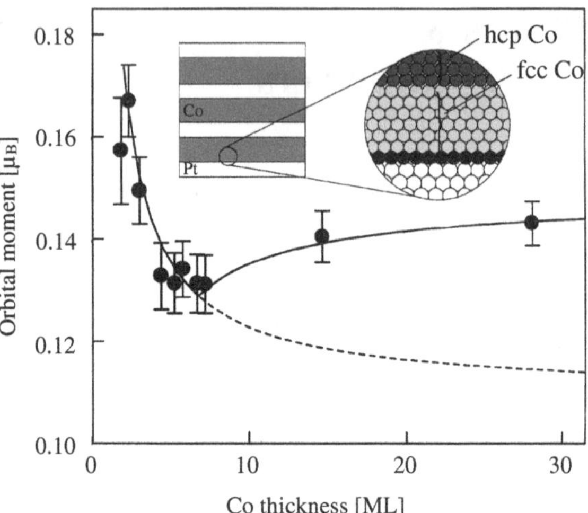

Fig. 3.1 Orbital magnetic moment as a function of the number of Co monolayers (ML) of [Co(t_{Co} ML)/Pt(7.5 ML)] multilayers. The *solid line* corresponds to a fit referring to the growth model presented in the inset. The *dashed line* denotes an extrapolation for the case of pure fcc Co growth [3]

The perpendicular Co orbital moment depending on the layer thickness is presented in Fig. 3.1. The strong increase of moment towards thinner Co layers is attributed to the strong interfacial 3d–5d hybridization between Co and Pt getting more influence. The gradual increase of orbital moment towards thicker Co layers is related to changes in the structure from fcc to hcp Co for film thicknesses greater than 6–8 ML, since the hcp structure possesses higher orbital moment. In general, the enhancement of orbital moment as well as the Co hcp structure causes PMA by spin-orbit coupling. Furthermore, a substantial Pt orbital moment induced by spin polarization was found also favoring PMA. In the manner of interfacial hybridization and spin polarization the roughness, intermixing, and possible alloy formation at the interface plays a crucial role for the magnetic anisotropy making the magnetic properties of Co/Pt multilayers very sensitive to their fabrication conditions [5, 7].

Beside the anisotropy induced by the interface of the multilayers, crystalline anisotropy provides another contribution to the effective anisotropy. With this regard Lin et al. [2] investigated the influence of the crystal orientation on the direction of the easy axis. Although the interfacial anisotropy points perpendicular to the film plane crystal orientation can force the easy axis to lie in the plane as shown in Fig. 3.2. Co/Pt multilayers grown epitaxial on a GaAs(111) single crystal along the Pt [110] direction exhibit an easy axis parallel to the film plane. This shows that the magnetocrystalline anisotropy can have a larger influence on the effective anisotropy than that resulting from the reduced symmetry at the interface. Therefore, crystalline orientation has to be taken into account for the fabrication of perpendicular Co/Pt multilayers. In principle for most purposes a polycrystalline Pt seed layer deposited at room temperature provides good growth conditions to obtain multilayers with PMA.

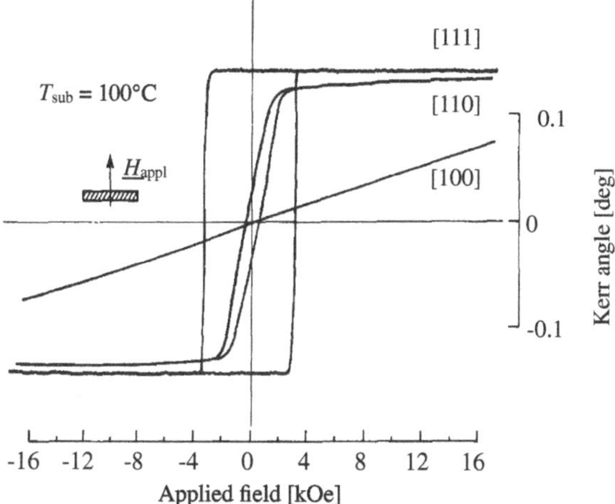

Fig. 3.2 Polar Kerr hysteresis loops of different [Co(3.7 Å)/Pt(16.8 Å)] multilayers grown epitaxial along three different directions. The Pt [111] orientation was grown on a GaAs(111) and the [110] as well as [100] orientation on a GaAs(100) single crystal [2]

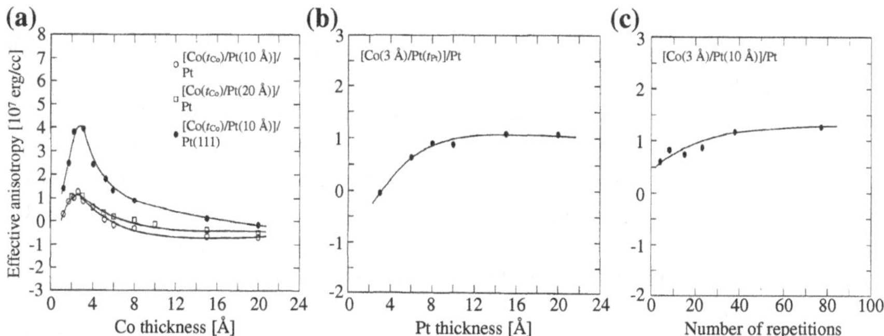

Fig. 3.3 Effective magnetic anisotropy of Co/Pt multilayers as a function of **a** the Co layer thickness, **b** the Pt layer thickness, and **c** the number of repetitions [2]. Please note, Pt and Pt(111) denote a polycrystalline and a textured seed layer, respectively

3.2 Magnetic Properties Depending on the Structure of the Multilayers

The present chapter is dedicated to some integral magnetic properties namely effective magnetic anisotropy and saturation magnetization of Co/Pt multilayers depending on their structural properties.

Figure 3.3 shows the effective anisotropy as a function of (a) the Co and (b) the Pt thickness as well as (c) the number of repetitions, respectively. The positive values of the anisotropy constant refer to an easy axis pointing out-of-plane. Concerning the thickness of the Co layer one observes a maximum anisotropy in the range from 2 to

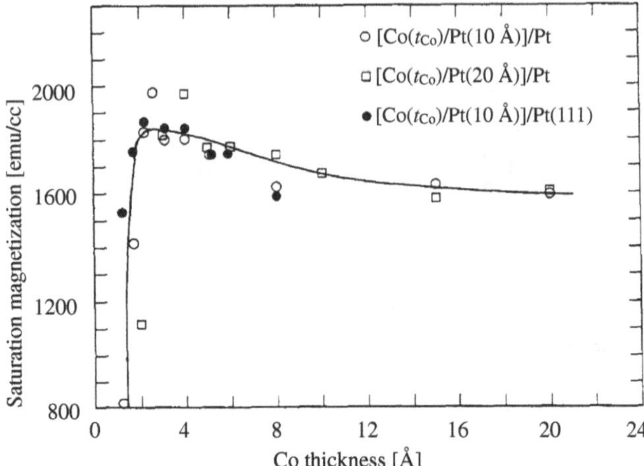

Fig. 3.4 Effective saturation magnetization of Co/Pt multilayers depending on the Co layer thickness [2]. Please note, Pt and Pt(111) denote a polycrystalline and a textured seed layer, respectively

4 Å, which is attributed to the greater influence of the interfacial 3d–5d hybridization as discussed in the previous chapter. In contrast to this the anisotropy increases towards thicker Pt layers and reaches its saturation value with thicknesses greater than 8 Å. An explanation for this behavior was given by Lin et al. [2] as the lack of exchange coupling between adjacent Co layers for thicker Pt layers. Furthermore, a decreasing coercivity for Pt layers thicker than 10 Å was observed. Similar to the Pt thickness an increasing number of repetitions leads to higher anisotropy values saturating in the range of 40 repetitions, which is related to the improving (111) texture with more repetitions.

The saturation magnetization as outlined in Fig. 3.4 depends strongly on the Co layer thickness. The effective value was obtained from the ratio of the total magnetization of the multilayer and the Co volume. For Co thicknesses less than 2 Å the effective value decreases below the value for bulk Co (1420 emu/cc) due to substantial decrease in Curie temperature. Multilayers with thicker Co films possess an up to 30 % higher saturation magnetization, which is attributed to the spin polarized Pt at the Co interfaces [16, 17].

Based on the presented and known magnetic properties Co/Pt multilayers with layer thicknesses of 4 Å for Co and 8 Å for Pt were used for all investigations concerning the interfacial exchange coupling in Fe–Tb/[Co/Pt] heterostructures, since these thicknesses provide sufficient high PMA and saturation magnetization in the multilayers.

References

1. P.F. Carcia, J. Appl. Phys. **63**, 5066 (1988)
2. C. Lin, G.L. Gorman, C. Lee, R.F.C. Farrow, E.E. Marinero, H.V. Do, H. Notarys, J. Magn. Magn. Mater. **93**, 194 (1991)
3. N. Nakajima, T. Koide, T. Shidara, H. Miyauchi, H. Fukutani, A. Fujimori, K. Iio, T. Katayama, M. Nývlt, Y. Suzuki, Phys. Rev. Lett. **81**, 5229 (1998)
4. J. Huang, M. Chen, C. Lee, T. Wu, J. Wu, C. Fu, J. Magn. Magn. Mater. **239**, 326 (2002)
5. I.B. Chung, Y.M. Koo, J.M. Lee, J. Appl. Phys. **87**, 4205 (2000)
6. S. Hashimoto, Y. Ochiai, K. Aso, J. Appl. Phys. **66**, 4909 (1989)
7. G.A. Bertero, R. Sinclair, C.H. Park, Z.X. Shen, J. Appl. Phys. **77**, 3953 (1995)
8. J. Huang, C. Lee, K.L. Yu, J. Appl. Phys. **89**, 7059 (2001)
9. S.J. Greaves, A.K. Petford-Long, Y.H. Kim, R.J. Pollard, P.J. Grundy, J.P. Jakubovics, J. Magn. Magn. Mater. **113**, 63 (1992)
10. H.J.G. Draaisma, W.J. M.de Jonge, F.J.A. den Broeder. J. Magn. Magn. Mater. **66**, 351 (1987)
11. T. Thomson, L. Abelmann, H. Groenland, *Magnetic Nanostructures in Modern Technology: Spintronics, Magnetic MEMS and Recording* (Springer, Berlin, 2008)
12. Y. Kawada, Y. Ueno, K. Shibata, IEEE Trans. Magn. **40**, 2489 (2004)
13. K. Takano, G. Zeltzer, D.K. Weller, E.E. Fullerton, J. Appl. Phys. **87**, 6364 (2000)
14. W. Peng, O. Keitel, R.H. Victoria, E. Koparal, J.H. Judy, IEEE Trans. Magn. **36**, 2390 (2000)
15. H. Ohmori, A. Maesaka, IEEE Trans. Magn. **36**, 2384 (2000)
16. T.R. McGuire, J.A. Aboaf, E. Klokholm, J. Appl. Phys. **55**, 1951 (1984)
17. G. Schütz, R. Wienke, W. Wilhelm, W.B. Zeper, H. Ebert, K. Spörl, J. Appl. Phys. **67**, 4456 (1990)

Chapter 4
Exchange-Bias Effect in F/FI Rare-Earth-Transition-Metal Heterostructures

In its common way the exchange-bias (EB) effect appears in F/AF bilayers, if the antiferromagnet is cooled below its blocking temperature in the presence of an external magnetic field and an exchange field produced by the magnetic moments of the ferromagnet, as originally found in the CoO/Co system by Meiklejohn and Bean in 1956 [1]. A schematic drawing of the effect with the two magnetic configurations situated above and below the ordering temperature of the AF layer is presented in Fig. 4.1.

Generally, one considers the frozen or pinned uncompensated spins at the interface of the antiferromagnet as reason for the biasing effect [2–4]. However, this explanation loses its validity in F/FI and FI/FI bilayer systems, which can also possess an exchange anisotropy [5–12]. Especially heterostructures with PMA, in which a FI amorphous rare-earth-transition-metal alloy like Fe–Tb is in the proximity of a transition-metal magnet, exhibit peculiar coupling interactions at the interface [13]. The magnetic coupling in such systems consists of two types of pair interactions, an antiparallel exchange between the transition-metal (RE) and transition-metal (TM) moments and a parallel exchange of the TM moments themselves [14]. Therefore, the EB effect can arise by both interactions simultaneously and occurs even for fully compensated interfaces. Furthermore, contrary to crystalline exchange coupled F/AF heterostructures where grain boundaries, atomic steps, or displacements reduce the exchange interaction at the interface and the maximum possible EB field, heterostructures based on FI systems with amorphous structure possess non of these morphology dependent reductions, which may lead to ultra high exchange fields. A schematics concerning the morphology influence on the interfacial exchange is given in Fig. 4.2.

One of the first observations about the EB effect in F/FI RE-TM heterostructures were made on TbCo/NiFe heterostructures [5–7]. It was found that the combination of NiFe with in-plane and amorphous TbCo with out-of-plane magnetic anisotropy produces a particular spin structure at the interface. The in-plane NiFe magnetization induces an in-plane component into the TbCo layer in the vicinity of the interface due to the strong exchange coupling. However, with increasing distance from the interface the TbCo layer relaxes back into its preferred out-of-plane magnetization

C. Schubert, *Magnetic Order and Coupling Phenomena*, Springer Theses,
DOI: 10.1007/978-3-319-07106-0_4,

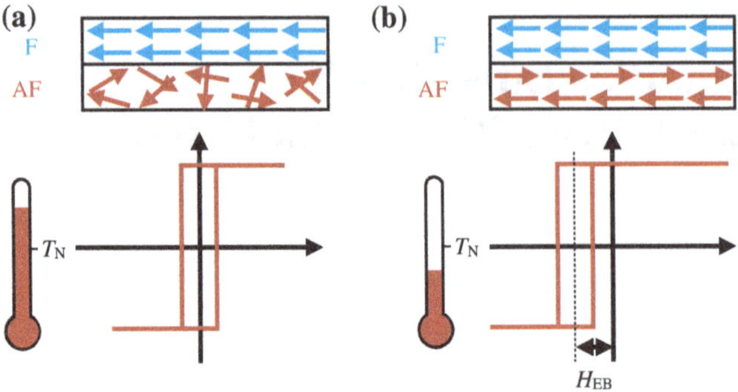

Fig. 4.1 Phenomenological schematics of the EB effect in F/AF bilayers. **a** Above the Néel temperature T_N of the AF layer no magnetic ordering exists and the F layer remains rather unaffected in its magnetization reversal illustrated by the symmetric hysteresis loop. **b** Below the ordering temperature of the AF layer strong interfacial exchange coupling sets in leading to an unidirectional anisotropy in the F layer with only one favorable remanent state. The resulting loop shift is denoted as EB field H_{EB}

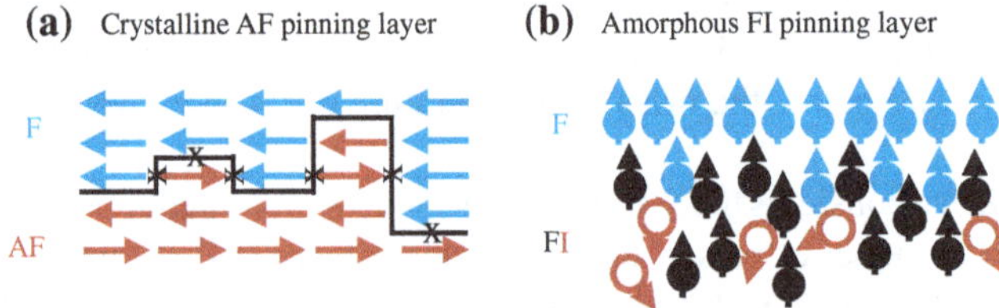

Fig. 4.2 Schematic drawing of the exchange coupling depending on the morphology at the interface of EB heterostructures. **a** Conventional crystalline F/AF EB bilayer stack. Grain boundaries, atomic steps and displacements reduce the maximum achievable exchange field. **b** EB system based on an amorphous FI pinning layer. Simultaneous exchange coupling between the magnetic sublattices of the ferrimagnet and the ferromagnet guarantees a strong exchange field. Even at the compensation point of the FI layer a full EB effect is possible. Furthermore, morphology dependent lowering of the exchange coupling is not existent

state. This particular configuration provides an exchange field of about 170 Oe. With regard to this phenomenological description micromagnetic simulations have shown that strong interfacial exchange between the NiFe and TbCo layer produce an in-plane uniaxial surface anisotropy at the interface, which relaxes into out-of-plane direction at the top of the TbCo layer by formation of 180° Bloch walls. As previously proposed by Mauri et al. [15] the formation and relaxation of the domain walls provide the driving force for the exchange anisotropy observed in this system. Furthermore, also pure in-plane exchange coupled F/FI heterostructures reveal the formation of 180° Bloch walls in the inappropriate F state between RE and TM moments as observed in GdFe/FeSn bilayers [9]. The profile of the orientational configuration of the magnetic moments in the exchange coupled heterostructure for different applied magnetic fields is outlined in Fig. 4.3.

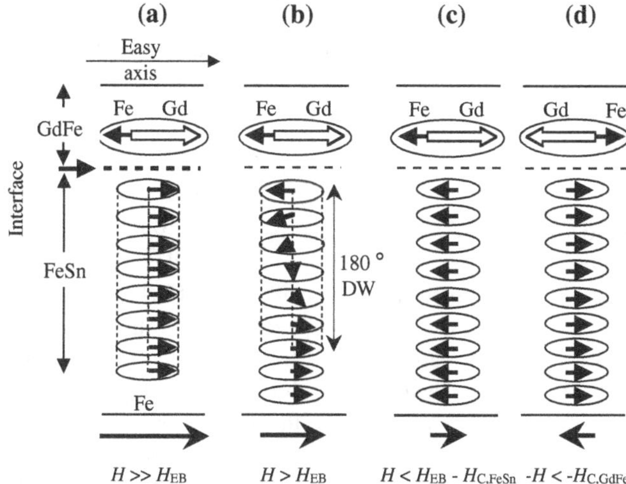

Fig. 4.3 Orientational moment configuration of GdFe/FeSn heterostructures for different applied magnetic fields. **a** A large external field overcomes the exchange field H_{EB} of the heterostructure aligning the Fe moments of the FeSn layer and the dominant Gd moment of the GdFe layer in one direction. The parallel oriented Fe and Gd moments produce a strong frustration at the interface region. **b** The external field in the vicinity of H_{EB} allows the nucleation of 180° Bloch walls providing a reduction of the interface frustration. **c** Small magnetic fields near zero give rise for a full reversal of the FeSn layer into the preferred antiparallel configuration between the Fe and Gd moments. **d** A rather small negative field reverses the whole layer stack due to the dominant Gd moments in the GdFe layer [9]

In the simplest approximation the EB field arises from the competition between the interfacial exchange coupling and the Zeeman energy [3, 16–18]:

$$H_{EB} = -\frac{J_{EB}}{M_F t_F}, \tag{4.1}$$

while J_{EB}, M_F, and t_F represent the interfacial exchange energy density, the saturation magnetization, and the thickness of the biased F layer, respectively. This relation leads to a strong overestimation of the EB field as the particular interfacial spin structure is not taken into account. Following the model of Mauri et al. [15] the exchange field of RE-TM heterostructures is directly related to the onset of 180° Bloch walls located at the interface region. The formation of such an interfacial domain wall (IDW) avoids the breaking of the exchange coupling, which allows an energy reduction in the system. In this manner the energy consumption of the domain wall σ_{IDW} determines the EB field:

$$H_{EB} = -\frac{\sigma_{IDW}}{2M_F t_F}. \tag{4.2}$$

And with regard to the domain wall energy $\sigma_{\mathrm{IDW}} \approx \sqrt{A_{\mathrm{eff}} K_{\mathrm{eff}}}$ the EB field is related to the effective exchange stiffness A_{eff} and magnetic anisotropy K_{eff} of the material at the interface region:

$$H_{\mathrm{EB}} \approx -\frac{\sqrt{A_{\mathrm{eff}} K_{\mathrm{eff}}}}{2 M_{\mathrm{F}} t_{\mathrm{F}}}. \tag{4.3}$$

The formation of an IDW as underlying mechanism for exchange anisotropy in F/FI RE-TM heterostructures takes also place in complete perpendicular composite systems. First observations were made in TbFeCo/[Co/Pt]$_{15}$ and [Pt/Co]$_{50}$/TbFe bilayers [13, 19]. In case of exchange coupled bilayers with out-of-plane easy axis of the magnetization, the formation of 180° Bloch walls in the inappropriate magnetization state is not possible. Nevertheless, the existence of an IDW was proofed by polarized neutron reflectometry measuring the depth profile of in-plane magnetization components. As result a "Néel-like" domain wall configuration was assumed. However, a comprehensive investigation of the correlation between the formation of an IDW and the onset of the EB effect in perpendicular heterostructures with respect to the intrinsic structural and magnetic properties of the amorphous RE-TM alloy film (i.e. anisotropy in the chemical short range order and sperimagnetism) is missing. In the context of this work exchange coupled heterostructures consisting of Co/Pt multilayers and amorphous Fe–Tb alloy films were extensively analyzed with regard to the magnetization reversal mechanism and the formation of an IDW depending on the temperature, stoichiometry, magnetic anisotropy, film thickness, and interfacial exchange coupling.

References

1. W.H. Meiklejohn, C.P. Bean, Phys. Rev. **102**, 1413 (1956)
2. S. Roy, M. Fitzsimmons, S. Park, M. Dorn, O. Petracic, I. Roshchin, Z. Li, X. Batlle, R. Morales, A. Misra et al., Phys. Rev. Lett. **95**, 47201 (2005)
3. F. Radu, H. Zabel, Springer Tracts Mod. Phys. **227**, 97 (2008)
4. K. O'Grady, L. Fernandez-Outon, G. Vallejo-Fernandez, J. Magn. Magn. Mater. **322**, 883 (2009)
5. W.C. Cain, M. Kryder, J. Appl. Phys. **67**, 5722 (1990)
6. N. Smith, W.C. Cain, J. Appl. Phys. **69**, 2471 (1991)
7. P.P. Freitas, J.L. Leal, L.V. Melo, N.J. Oliveira, L. Rodrigues, A.T. Sousa, Appl. Phys. Lett. **65**, 493 (1994)
8. R. Sbiaa, H. Le Gall, Appl. Phys. Lett. **75**, 256 (1999)
9. F. Canet, S. Mangin, C. Bellouard, M. Piecuch, Europhys. Lett. **52**, 594 (2000)
10. F. Canet, S. Mangin, C. Bellouard, M. Piecuch, A. Schuhl, J. Appl. Phys. **89**, 6916 (2001)
11. S. Mangin, F. Montaigne, A. Schuhl, Phys. Rev. B **68**, 140404 (2003)
12. F. Radu, R. Abrudan, I. Radu, D. Schmitz, H. Zabel, Nat. Commun. **3**, 715 (2012)
13. S. Mangin, T. Hauet, P. Fischer, D. Kim, J.B. Kortright, K. Chesnel, E. Arenholz, E.E. Fullerton, Phys. Rev. B **78**, 024424 (2008)
14. I. Campbell, J. Phys. F Met. Phys. **2**, L47 (1972)
15. D. Mauri, H.C. Siegmann, P.S. Bagus, E. Kay, J. Appl. Phys. **62**, 3047 (1987)

16. R. Stamps, J. Phys. D Appl. Phys. **33**, R247 (2000)
17. J. Nogués, I. Schuller, J. Magn. Magn. Mater. **192**, 203 (1999)
18. W.H. Meiklejohn, J. Appl. Phys. **33**, 1328 (2009)
19. S.M. Watson, T. Hauet, J.A. Borchers, S. Mangin, E.E. Fullerton, Appl. Phys. Lett. **92**, 202507 (2008)

Chapter 5
Experimental Techniques

The preparation and analysis tools for the investigation of the structural and magnetic properties of flat and patterned Fe–Tb alloy films, Co/Pt multilayers, and exchange coupled heterostructures within the purpose of this work will be briefly presented in this chapter.

5.1 Film Deposition Using Magnetron (Co-)Sputtering from Element Targets

The collision between high kinetic ions and the surface of a solid can lead to evaporation of atoms and molecules as well as cluster particles. The accelerated ions migrate into the solid and yield energy to the lattice atoms by elastic-collisions. During this cascades a certain amount of atoms receives a refused momentum and enough energy to evaporate from the solid. This process, named as sputtering, is applicable for technical reason to deposits thin homogenous films in the nm-size-regime.

A principle scheme of a magnetron sputter gun is drawn in Fig. 5.1. The sputtering process requires vacuum to ensure sufficient high mean free path for the ions and electrons and to avoid arc discharge between the anode and the target. The necessary ions are provided by the supply of highly pure gases like Ar, Xe, O_2, etc. regulated to partial pressures in the order of μbar. Generally, a portion of the gas atoms is always ionized. If an electric field $\underline{E}$ is applied these ions move to the target, while the free electrons become accelerated to the anode. This motion of charged particles causes collision ionization and leads to plasma formation. In case of magnetron sputtering an overlaying magnetic field $\underline{H}$ forces the electrons to cycloidal movements, which limits the plasma to an area located above the target. Thus, the plasma has no direct contact with the target. Due to the radial symmetry of the applied fields presented in the scheme of Fig. 5.1 a plasmatorus occurs. The ions, which are not limited to this torus, hit the target and sputter material from it. The decruited atoms, molecules or clusters hold no charge and leave the magnetron sputter gun straight-lined as far as

C. Schubert, *Magnetic Order and Coupling Phenomena*, Springer Theses,
DOI: 10.1007/978-3-319-07106-0_5,

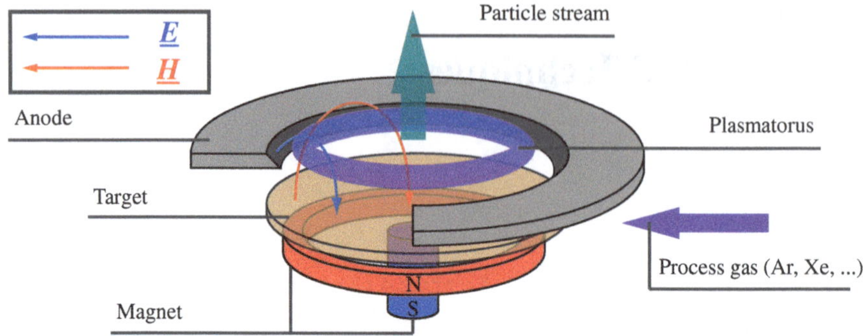

Fig. 5.1 Schematic drawing of the components of a magnetron sputter gun. The crossing electrical and magnetic fields are illustrated as *blue* and *red* arrows, respectively. The purple ring represents the plasmatorus built by the electrons and ionized process gas atoms

they are deposited at the surface of a substrate or at the wall of the vacuum chamber. The spacial distribution of the particle stream is inhomogeneous and exhibits a club like shape in the simplest approximation. Nevertheless homogeneous thin films can be achieved by adjusting the sample holder slightly off the focus of the sputtering club and rotating the substrate during deposition.

The preparation of Fe–Tb alloy films and Co/Pt multilayers was realized in two similar ultra high vacuum (UHV) chambers by (co-)sputtering from element targets with purities higher than 99.96 %. The first one is a commercial system from *Bestec*, which was used in cooperation with Florin Radu at the Helmholtz-Zentrum Berlin (HZB). The second system is a self-made combined sputtering and molecular beam epitaxy (MBE) chamber labeled as *SMBE*. A schematic drawing and some photos of the system are shown in Fig. 5.2. Both chambers exhibit a base pressure better than 5×10^{-8} mbar and an adaptable Ar working pressure between 1.5 and 9 μbar due to similarities only the second system will be described in detail. Information about the *Bestec* system can be deduced from the company website [1]. The self-made system possess a two pocket e-beam evaporation gun and a fixed sample holder in the MBE chamber and four focused magnetron sputter guns with a rotatable and vertical moveable sample holder in the sputtering chamber. The whole system becomes evacuated by a roots and turbo molecular pump. Furthermore, the MBE part is pumped by an ion getter pump and the loadlock has an additional turbo pump. Please note that for the sample preparation only the sputtering chamber was used.

The regulation of the Ar pressure during the deposition is realized by a flow controller, which is connected to a gas bottle of Ar with 99.999 % of purity. A quartz balance measures the sputtering rates before the deposition to calibrate the electric power and time with regard to the desired film thickness. The crystal is located near to the sample holder. The small deviation between both positions are compensated by tooling factors. These tooling factors are estimated for each magnetron gun from the ratio between the nominal and real thickness of corresponding reference films measured by X-ray reflectometry. During the deposition the sample holder is placed a

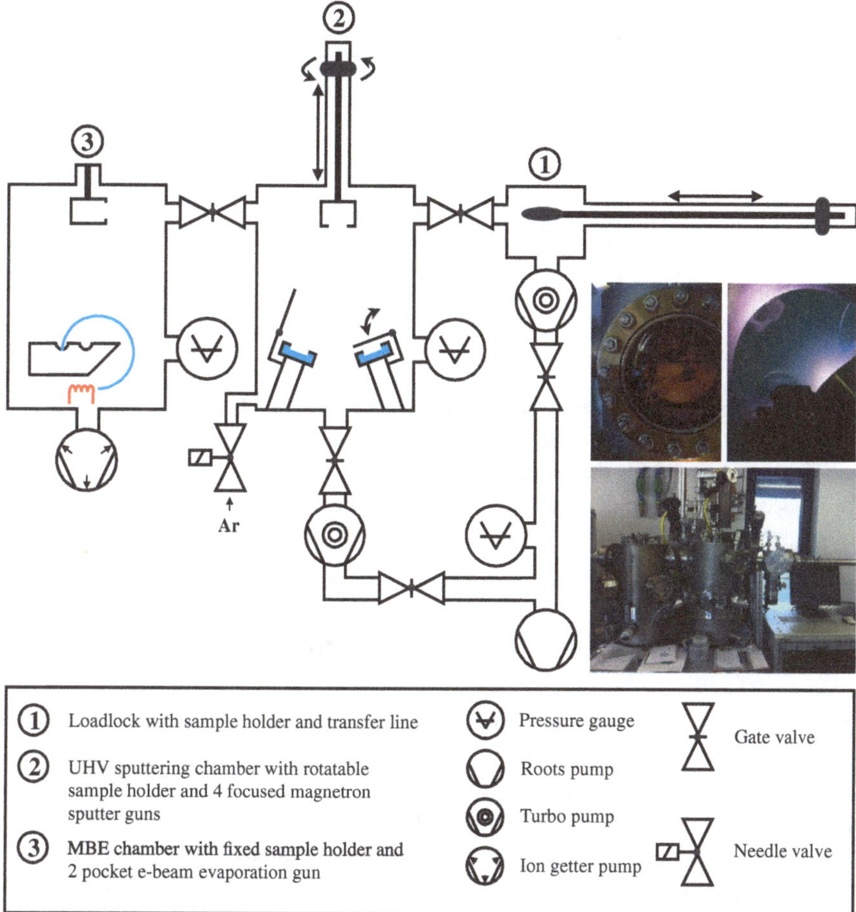

Fig. 5.2 Schematic drawing of the combined sputtering and MBE UHV chamber labeled as *SMBE*. Four focused magnetron sputter guns permit up to four component alloy film as well as multilayer deposition. The two pocket e-beam evaporation gun can be used for epitaxial growth of single layers. A photo of the whole system, the view port to the sample holder, and a view into the chamber during deposition are placed on the right side of the drawing

bit below the focus point of the magnetrons and rotates to provide good homogeneity. The chamber has no extra sample shutter. The start and stop of the deposition process is controlled by the target shutters above the magnetrons, which are operated by computer.

As it was already mentioned, thin $Fe_{100-x}Tb_x$ alloy films were prepared using the co-sputtering technique. The composition of the binary alloy is controlled by adjusting the electric power of the magnetron discharges to obtain the right ratio of the deposition rates Φ. The deposition rate is defined as film thickness t divided by the deposition time τ. Expressing the thickness via the mass m, the density ρ, and area S of the deposited material to $t = m/\rho S$ the ratio between the deposition rates of the Fe and Tb target is given by the following equation:

$$\frac{\Phi_{\mathrm{Fe}}}{\Phi_{\mathrm{Tb}}} = \frac{m_{\mathrm{Fe}}}{m_{\mathrm{Tb}}} \cdot \frac{\rho_{\mathrm{Tb}}}{\rho_{\mathrm{Fe}}}. \tag{5.1}$$

Taking into account that the mass ratio is related to the atomic numbers A_{Fe} and A_{Tb} of the elements by $\frac{m_{\mathrm{Fe}}}{m_{\mathrm{Tb}}} = \frac{100-x}{x} \cdot \frac{A_{\mathrm{Fe}}}{A_{\mathrm{Tb}}}$ the relation for the ratio of the deposition rates results to:

$$\frac{\Phi_{\mathrm{Fe}}}{\Phi_{\mathrm{Tb}}} = \frac{100-x}{x} \cdot \frac{A_{\mathrm{Fe}}}{A_{\mathrm{Tb}}} \cdot \frac{\rho_{\mathrm{Tb}}}{\rho_{\mathrm{Fe}}}. \tag{5.2}$$

As an example, a ratio between the deposition rates of 1.467 for $Fe_{80}Tb_{20}$ and 0.856 for $Fe_{70}Tb_{30}$ can be deduced from the equation above. Please note that for all Fe–Tb alloy depositions the Fe rate was kept constant at 0.6 Å/s and the Tb rate was varied according to the desired stoichiometry, which was verified using Rutherford backscattering spectrometry.

5.2 The Fabrication of Nanodot Arrays by Pre-patterning

The pre-patterned substrates were produced by nanoimprint lithography [2–4] (NIL) where the structure of the imprinted resist is transferred via chemical etching into a Ta layer, which serves then as hard mask to bring the preferred structure into the silicon wafer using a plasma etch step. In detail, the stamp for the imprint is fabricated by electron beam lithography [5]. As resist polymethylmethacrylat (PMMA) is used. After the deposition of a 10 nm to 40 nm thick Ta layer, a PMMA film is spin coated on top in which the positive structure is pressed by the stamp. Then the Ta hard mask is produced by reactive etching. And in a subsequent Ar sputter etching process the preferred structure becomes transferred into the substrate due to the erosion of the sacrificial Ta mask. Please note that the thickness of the hard mask and the time of the etching needs to be adjusted to the desired height of the patterning.

By NIL arrays of silicon nanodots with a diameter of 30 nm, a height of about 16 nm, and a period of 60 nm were created in high quality over a 1 mm broad ring on a complete 3-inch wafer. The scanning electron microscopy image in Fig. 5.3 shows the top view of the nanodots. Please note that their arrangement deviates from a hexagonal lattice.

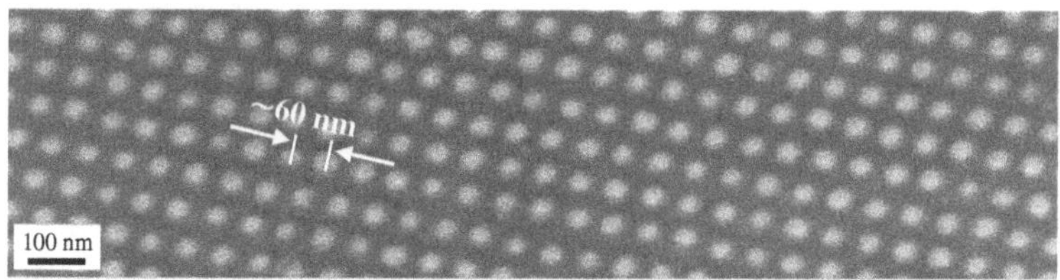

Fig. 5.3 Scanning electron microscopy image of a pre-patterned Si wafer with natural oxide presenting a nanodot array with 30 nm dot size and 60 nm period

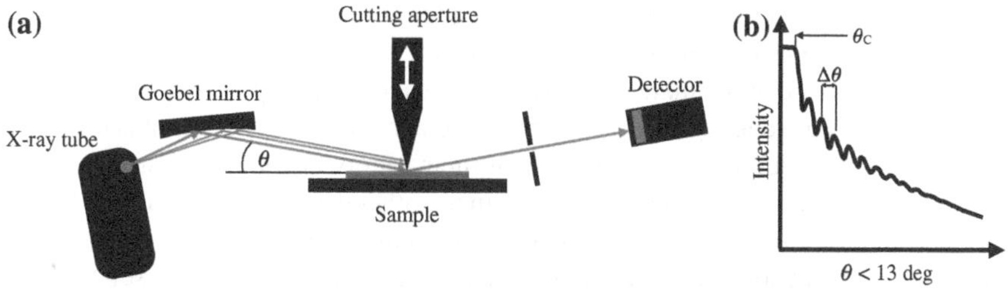

Fig. 5.4 **a** Schematic drawing with beam geometry of a XRR setup. The Goebel mirror and cutting aperture provide a parallel beam and a well defined footprint onto the sample surface in this small angle geometry. **b** Typical XRR spectra for small angles θ with Kiessig fringes

5.3 Structural Characterization Techniques

5.3.1 Estimation of Film Thickness by X-Ray Reflectometry

The nominal thickness of an evaporated film, obtained from the frequency shift of the quartz balance during the sputtering process, differs from the real thickness value due to a deviation between the position of the sample holder and the quartz balance in the vacuum chamber. As mentioned above, this difference can be compensated by a tooling factor, which takes the sputtering geometry and sample holder position into account. In order to estimate the tooling factor and calibrate the thickness measurement the real thickness of a deposited sample film needs to be analyzed. One of the most suitable techniques for thickness measurements is X-ray reflectometry (XRR).

Thin films, which are illuminated by X-rays under small angles θ as illustrated in the scheme of Fig. 5.4a, provide a total reflection of the incident light as far as a critical angle θ_C is not exceeded. For larger angles the reflected intensity decreases rapidly. This decay is superimposed by oscillations of the intensity; the so called Kiessig fringes (see Fig. 5.4b). These oscillations possess a characteristic period $\Delta\theta$ related to the total thickness of the thin film. Taking into account the wave length λ of the incident light, the thickness t can be calculated with the following equation [6]:

$$t = \frac{\lambda}{2 \sin \Delta\theta}. \tag{5.3}$$

The estimation of the layer thickness in order to calibrate the sputter deposition system in the *SMBE* chamber was realized by XRR using a four circle diffractometer *XRD 3000PTS* from *Seifert*.

5.3.2 Stoichometry Measurements by Rutherford Backscattering Spectrometry

High kinetic ions can penetrate solids and scatter at the atoms of the bulk material. Depending on the kinetic energy of the ions and the scattering cross-section as well as the crystal structure of the material two different scenarios are possible. If the ions do not possess enough energy or the scattering cross-section of the material is too large, they will be absorbed by the solid as all of their energy becomes transferred to the lattice atoms during the scattering process. In the other case ions with enough kinetic energy can leave the material but will be scattered in different angles with respect to their incident direction. In particular, the energy and intensity of these backscattered ions yield information about the stoichiometry of the scattering material. This is utilized in the Rutherford backscattering spectrometry (RBS) technique.

In first approximation the backscattering can be described as classical elastic interactions between the energy-rich ions and the static atoms localized in the lattice of the sample material. Thus, the kinetic energy E_{Ion} of the backscattered ions depends on their incident energy $E_{0,\text{Ion}}$ and mass m_{Ion} as well as the mass of the scattering nuclei m_{Nucleus} [7]:

$$E_{\text{Ion}} = E_{0,\text{Ion}} \left(\frac{m_{\text{Ion}} \cdot \cos\vartheta + \sqrt{m_{\text{Nucleus}}^2 - m_{\text{Ion}}^2 \sin^2\vartheta}}{m_{\text{Ion}} + m_{\text{Nucleus}}} \right)^2, \tag{5.4}$$

where ϑ is the angle of the backscattered ions with respect to their incident direction.

A schematic of the setup for RBS is given in Fig. 5.5a. A linear accelerator provides high-kinetic ions (often H^+ or He^{++}) with energies between 1 and 3 MeV. The accelerated ions hit the surface of the sample and become backscattered. The detector is mounted in a fixed angle ϑ towards the incident direction and detects the arriving ions with respect to their number and energy. A qualitative drawing of the energy distribution of these ions is shown in Fig. 5.5b. The plateau is caused by the substrate of the sample. The narrow peak comes from the thin film on top of the substrate and provides information about the chemical element the film consists off. In alloy films more than one peak is observed giving rise to identify the included elements and their composition due to the different integral intensities. However, a correction is necessary since the intensity depends on the differential cross-section of the scattering process, which scales with the square of the atomic number of the scattering element [7].

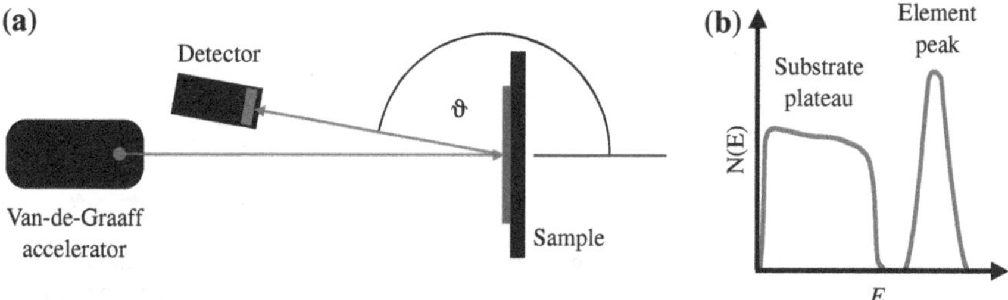

Fig. 5.5 **a** Schematic drawing with beam geometry of a RBS setup. The Van-de-Graaff accelerator provides high-kinetic H^+ or He^{++} ions focused onto the sample. The detector with multi channel analyzer sits in an angle ϑ before the specimen. **b** Typical energy distribution for RBS with a substrate plateau and element peak

Rutherford backscattering measurements of the thin Fe–Tb alloy films were done in cooperation with the Helmholtz-Zentrum Dresden-Rossendorf. For the scattering He^{++} ions with an energy of 1.7 MeV were used. The high-energy was provided by an Van-de-Graaff accelerator. The backscattering angle was held constant at $\vartheta = 170\,\text{deg}$ and the energy distribution of the ions was obtained by a semiconductor detector with multi-channel analyzer. The stoichiometry calculations from the RBS spectra were performed by using the simulation software *SIMNRA* allowing an accuracy within 1 at.% [8].

5.3.3 Determining Crystal Structures by X-Ray Diffraction

X-ray diffraction (XRD) is a suitable technique to determine the crystalline orientation of thin films. Atoms in condensed matter act as local scattering centers for incident X-rays and generate interfering elementary waves. Depending on the crystalline structure constructive interference can occur for certain incident angles (glancing angles). This is observed macroscopically as a reflection of the X-rays. The condition for constructive interference is given by the Laue formalism (see the geometric expression in Fig. 5.6a demanding the difference between the scattered and incident wave vector $(\underline{k}' - \underline{k})$ to be an integer multiple of the reciprocal lattice vector $\underline{R}$ [9]:

$$e^{i[(\underline{k}'-\underline{k})\cdot\underline{R}]} = 1. \tag{5.5}$$

Transferring the Laue formalism into real space leads to the Bragg equation, which gives the correlation between the glancing angle θ_{hkl}, the wave length of the incident light λ, and the distance of the lattice planes a_{hkl}:

$$a_{\text{hkl}} = \frac{N\lambda}{2\sin\theta_{\text{hkl}}}, \tag{5.6}$$

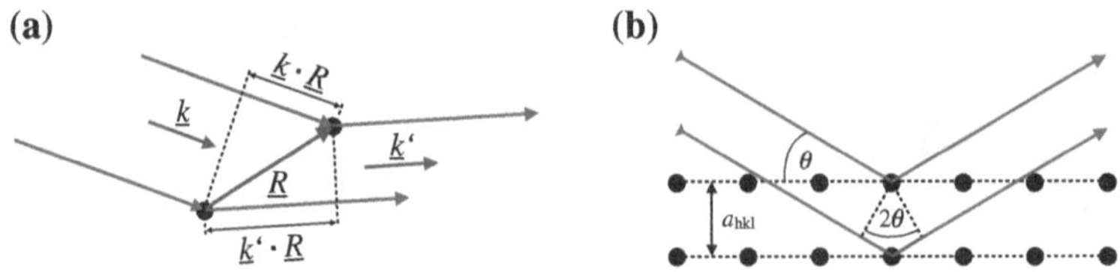

Fig. 5.6 **a** Laue condition in reciprocal space with incoming and outgoing wave vectors $\underline{k}$ and $\underline{k}'$ and reciprocal vector $\underline{R}$. **b** Bragg condition in real space with diffraction angle θ and the distance of two lattice planes a_{hkl}

where N is an integer number. In this manner constructive interference appears if the path difference of two X-rays scattered at different lattice planes is an integer multiple of the X-ray wavelength. In order to determine the crystal structure and orientation the $\theta/2\theta$-method is commonly used. The schematic in Fig. 5.6b illustrates the measuring geometry of this method. By varying the incident angle θ the obtained Bragg reflexes provide distinctive information about crystal structure and orientation of the thin film. In particular the full width at half maximum (FWHM) $\Delta\theta_{\text{hkl}}$ of the Bragg peak allows an estimation of the minimum cross-sectional coherence length and with it the size $d_{\text{c},\perp}$ of the corresponding grains according to the Scherrer equation [9]:

$$d_{\text{c},\perp} = 2\sqrt{\ln 2/\pi} \cdot \frac{\lambda}{\Delta\theta_{\text{hkl}} \cos\theta_{\text{hkl}}}. \tag{5.7}$$

Further information can be deduced from rocking curve measurements where the detector is fixed to the centroid position $2\theta_{\text{hkl}}$ of the Bragg peak while the sample is tilted by the angle ω with respect to the X-ray source and detector. These so called ω-scans allow to obtain information about the tilt and size of the crystallites. Beside the orientational distribution of the crystallites the FWHM $\Delta\omega_{\text{hkl}}$ of the rocking curve provides a measure of the minimum lateral grain size $d_{\text{c},\parallel}$ following the modified Scherrer equation for the rocking curve [10]:

$$d_{\text{c},\parallel} = 2\sqrt{\ln 2/\pi} \cdot \frac{\lambda}{\Delta\omega_{\text{hkl}} \sin\omega_{\text{hkl}}}. \tag{5.8}$$

The investigations concerning crystal structure and morphology of the Fe–Tb alloy films and seed layers were realized by XRD in a universal diffractometer *XRD-7* with Eulerian cradle and x-y-stage as well as a four circle diffractometer *XRD 3000PTS* from *Seifert*. The incident X-ray beam is provided by a Cu Kα_1 source with a wavelength of 1.5418 Å.

5.3.4 Film Morphology, Crystal Structure, and Topography Investigations by Electron Microscopy

Conventional light microscopy is limited in resolution given by the wavelength of the incident light. Structures smaller than the wavelength are not visible due to diffraction phenomena. The smallest visible distance d between two different objects can be calculated using Abbe equation $d = \lambda/n' \sin\beta$, where n' is the refractive index of the medium between the object and objective and β denotes the opening angle of the objective. With this a maximum resolution of about 200 nm is given.

For higher resolution accelerated electrons can be used. Due to the wave-particle duality each particle exhibits a so called de Broglie wavelength, which depends on its relative momentum $\lambda = h/p$ with h as the Planck's constant. Taking into account the relativistic correction due to high accelerated electrons the wavelength can be estimated using the following equation:

$$\lambda \approx \frac{h}{\sqrt{2m_\mathrm{e}E_\mathrm{kin}\left(1+\frac{2E_\mathrm{kin}}{m_\mathrm{e}c^2}\right)}}, \tag{5.9}$$

where c is the speed of light, m_e denotes the rest mass, and E_kin the kinetic energy of the electrons. Typical energies in common electron microscopes range between 100 and 300 keV leading to wavelengths between 3.3 and 1.5 pm, respectively. Based on the smallest wavelength a maximum resolution of 50 pm is possible [11].

In electron microscopes electrons are emitted from a filament cathode, accelerated through a high voltage anode, and focused by electrostatic and electromagnetic lenses onto a thin specimen. The high energy electrons interact with the atoms of the specimen and become scattered, diffracted or absorbed. These interactions lead to the emission of secondary electrons and characteristic X-rays giving rise for different imaging techniques.

Transmission electron microscopy (TEM) generates images of thin specimens from the scattering contrast of transmitted electrons projected on either an imaging plate or CCD camera. This technique allows the investigation of morphology, grain size, and crystal structure of thin films by dark and bright field imaging. In order to guaranty a sufficient high electron transparency the typical sample thickness need to be less than 100 nm for both lateral and cross-sectional imaging. This requires a sophisticated preparation procedure involving heating the sample, mechanical grinding, and Ar ion etching. The TEM investigations of single layers and heterostructures consisting of Fe–Tb alloy films and Co/Pt multilayers were realized in a *Phillips CM 20 FEG* microscope [12] with an accelerating voltage of 200 kV.

Besides morphology and crystal structure revealed from TEM investigation a surface image of the sample gives important information about the topography. One suitable technique to depict surface images is the scanning electron microscopy (SEM). In such a microscope a focused electron beam scans over the interesting area of a sample. Due to the incident electrons secondary electrons are emitted from the thin film in the range of a view nanometer below its surface and captured by a scintillation counter. In this manner the detected electrons yield direct information about the material in the surface region. The contrast in a SEM image originates from the interaction of the electrons with the local charge density as well as the grain boundaries of the crystallites. Furthermore, absorption contrast arises by shadowing effects due to the perspective of the scintillation counter. In the context of this work surface images of arrays of Fe–Tb nanodots were obtained by a *Nova NanoSEM 200* [13] from *FEI Company*.

5.3.5 Surface Imaging by Atomic Force Microscopy

Another technique dedicated for the investigation of surface structures and topography is the atomic force microscopy (AFM) [14]. A cantilever with an atomically sharp tip on its one end scans over the sample in the vicinity of its surface. The scanning is commonly realized by a combined stepper and piezo motor stage. Due to Van der Waals interactions between the atoms of the tip and the surface the cantilever becomes bended. The resulting curvature of the cantilever can be measured using a laser and a split diode detector as it is shown in the schematic drawing of Fig. 5.7. From the detection of the laser beam the force on the cantilever can be deduced taking into account its length and spring constant. With this it is possible to obtain a height profile from the sample.

A fast and contact free method to measure height profiles is the *Tapping Mode*™, where the cantilever becomes excited near its resonant frequency while scanning over the sample surface. Differences in height change the Van der Waals force acting on the cantilever, which leads to changes in its resonant frequency. This is accompanied by a registrable change of the laser amplitude allowing to image the surface topography. The imaging was realized with a *Dimension 3000* microscope from *Veeco instruments*.

The Van der Waals force is only one of the possible interactions between the tip and the sample surface. Beside adhesion, capillary and contact forces magnetic samples for instance generate dipolar interactions due to magnetic stray fields. This allows the imaging of these stray fields using tips coated with magnetic material [15]. Further information about this technique will be given later.

5.4 Investigation of Magnetic Properties

5.4.1 Magnetometry by Magneto-Optical Kerr Effect Measurements

The turning of light polarization after the reflection of an incident beam on a magnetic surface is referred to as magneto-optical Kerr effect (MOKE) named after J. Kerr, who discovered this effect in 1877 [16]. Due to its proportionality to the magnetization,

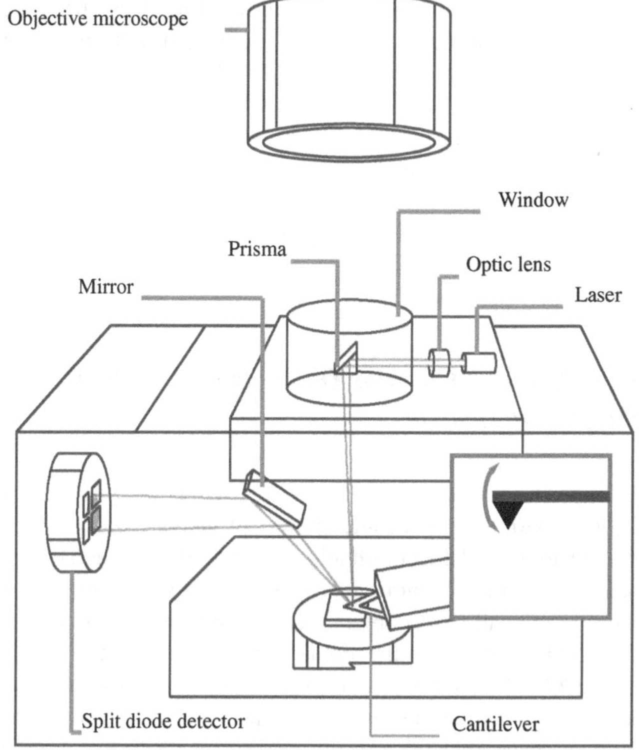

Fig. 5.7 Schematic drawing of the components of an atomic force microscope

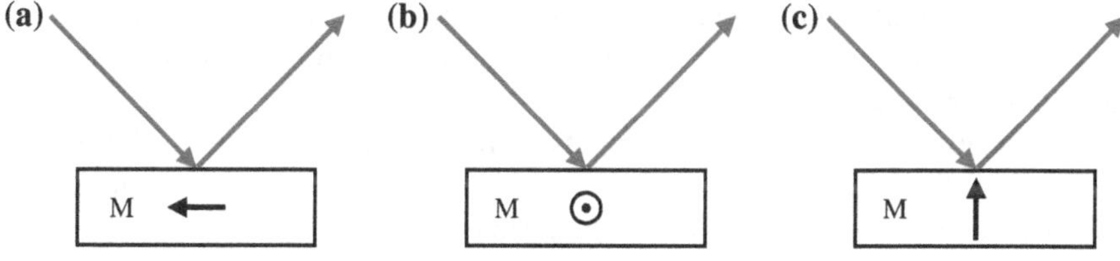

Fig. 5.8 Schematics of the geometries in which different MOKE signals can be obtained: **a** longitudinal, **b** transversal, and **c** polar geometry

high sensitivity, and simplicity the MOKE measurement technique is used for many different purposes, in particular the investigation of local magnetic properties in thin films is very common [17, 18]. Since absorption limits the penetration depth of the laser beam to less than 50 nm a substrate induced magnetic contribution will not influence the MOKE signal. Although MOKE measurements provide no absolute values of the magnetization they serve as a good pre-characterization tool to obtain hysteresis loops. In general the measurements can be realized in the longitudinal, transversal, and polar geometry as outlined in Fig. 5.8 below.

For standard polar MOKE measurements in order to cross-check the fabricated thin magnetic films a setup operating with a collimated 670 nm laser beam and a 20 kOe electromagnet was used in this work [19]. Furthermore, angular dependent remanence measurements applied on Fe–Tb flat films and nanodot arrays were performed in a rotatable polar setup with a focused 635 nm laser beam and a 10 kOe electromagnet [20]. In order to obtain remanence curves the samples were saturated in a positive field of 10 kOe followed by a negative reversal field. Afterwards the Kerr rotation was measured in zero field. This procedure was repeated for each reversal field point.

5.4.2 X-Ray Magnetic Circular Dichroism: An Element Specific Probe for Magnetism

Since its first experimental discovery in 1987 by Schütz et al. [21] X-ray magnetic circular dichroism (XMCD) has become a powerful analysis tool probing the spin and orbital moment of atoms and their orientation in the solid. Although magnetic circular dichroism has been known for more than hundred years [22] only the extension to the X-ray energy range allows to obtain element specific information owing to the characteristic binding energies of the atomic core electrons in the material.

In general the XMCD describes the difference in the absorption for right and left circularly polarized X-ray photons due to the different number of unoccupied spin-up and spin-down states above the Fermi level of magnetic materials. Particularly the absorption depends strongly on the relative orientation of the photons angular momentum $\pm\hbar$ and the magnetization. While diamagnetic and paramagnetic materials exhibit a significant effect only at very high magnetic fields 3d transition-metal [21, 23–25] and 4f transition-metal [26–30] ferromagnets possess strong dichroism already in zero field.

The absorption $\alpha(E)$ of X-ray photons with a certain energy in a solid is characterized by the logarithmic ratio of the incident $I_0(E)$ and the transmitted beam intensity $I(E)$ and is correlated to the energy dependent absorption coefficient $\mu(E)$ as well as the thickness t of the material:

$$\alpha(E) = \ln\left(\frac{I_0(E)}{I(E)}\right) = \mu(E) \cdot t. \tag{5.10}$$

Following the simplest approximation the photon absorption can be explained by the excitation of photo-electrons from the atomic core levels s, p, d, ... to higher orbitals. This leads to characteristic sharp steps in the absorption spectra labeled as K, L, M, ... corresponding to their originating core levels with the quantum numbers $n = 1, 2, 3, \ldots$, respectively. As an example the X-ray absorption by the excitation of 2p core electrons to unfilled 3d states is outlined in the simple one-electron picture in Fig. 5.9a. The L-edge X-ray absorption spectra are mainly determined by the electronic transition from the spin-orbit split 2p core shell $2p_{1/2}$ L_2 $(l - s)$ and $2p_{3/2}$

L_3 $(l+s)$ to empty conduction d-band states above the Fermi level. The transitions to s-band states are present but contribute 20 times weaker to the spectra [31]. Further split core levels are $M_{2,3}$ ($3p_{1/2}$, $3p_{3/2}$), $M_{4,5}$ ($3d_{1/2}$, $3d_{3/2}$), ... with the $M_{4,5}$-edges being suitable for the investigation of the 4f magnetism of rare-earth elements.

The dichroism in the X-ray absorption spectra originates from the angular momentum transfer of right or left circularly polarized photons with $\hbar$ or $-\hbar$ to the excited photoelectrons, respectively. The electrons coming from spin-orbit split levels like the $p_{1/2}$ and $p_{3/2}$ states become partially spin polarized due to the spin-orbit coupling. Because of the opposite coupling for the $2p_{1/2}$ and $2p_{3/2}$ levels ($l-s$ and $l+s$, respectively) the spin polarization occurs also opposite for the two absorption edges as it is illustrated in Fig. 5.9b. Depending on the spin polarization of the photoelectrons with respect to the number of unoccupied states in the spin-split valence shell different probabilities for the electronic transition exist. The quantization axis and with it the number of unoccupied spin-up and spin-down states depend on the orientation of the magnetization in the material. In particular the alignment of the magnetization parallel or antiparallel to the photon spin direction leads to a maximum or minimum transition probability for the polarized photoelectrons and therefore the X-ray absorption. The different absorption abilities are denoted with the absorption coefficient for right and left circularly polarized light $\mu^+(E)$ and $\mu^-(E)$, respectively. The XMCD spectra drawn in Fig. 5.9b, c arise from the difference in absorption $\Delta\mu(E) = \mu^+(E) - \mu^-(E)$. Based on the dichroic difference intensities A and B obtained from the integration of the XMCD spectra over the L_3- and L_2-edge the orbital and the spin moment can be calculated according to the sum rules $[A + B]$ and $[A - 2B]$, respectively [32, 33].

In general the z-component of the orbital moment $\langle L_z \rangle$ in units of μ_B/atom can be estimated using the following sum rule [32]:

$$\langle L_z \rangle = \frac{4}{3} \cdot \frac{\int_{j^+ + j^-} \mathrm{dE}\, \Delta\mu(E)}{\int_{j^+ + j^-} \mathrm{dE}\, (\mu^+(E) + \mu^-(E))} \cdot \frac{l(l+1)}{l(l+1) + 2 - l_c(l_c+1)} \cdot (4l + 2 - n_e), \tag{5.11}$$

where l_c and l denote the azimuthal quantum numbers for the core and valence level states, respectively, and n_e is the number of occupied states in the valence level. The symbols j^+ and j^- indicate the integration over the absorption edges of the two spin-orbit split levels. For the measurement of the orbital moment in transition-metal and rare-earth materials typically X-ray absorption is performed at the $L_{2,3}$- and $M_{4,5}$-edges according to these particular expressions:

$$\langle L_z \rangle = \frac{4}{3} \cdot \frac{\int_{L_3+L_2} \mathrm{dE}\, \Delta\mu(E)}{\int_{L_3+L_2} \mathrm{dE}\, (\mu^+(E) + \mu^-(E))} \cdot (10 - n_{3d}), \tag{5.12}$$

$$\langle L_z \rangle = 2 \cdot \frac{\int_{M_5+M_4} \mathrm{dE}\, \Delta\mu(E)}{\int_{M_5+M_4} \mathrm{dE}\, (\mu^+(E) + \mu^-(E))} \cdot (14 - n_{4f}). \tag{5.13}$$

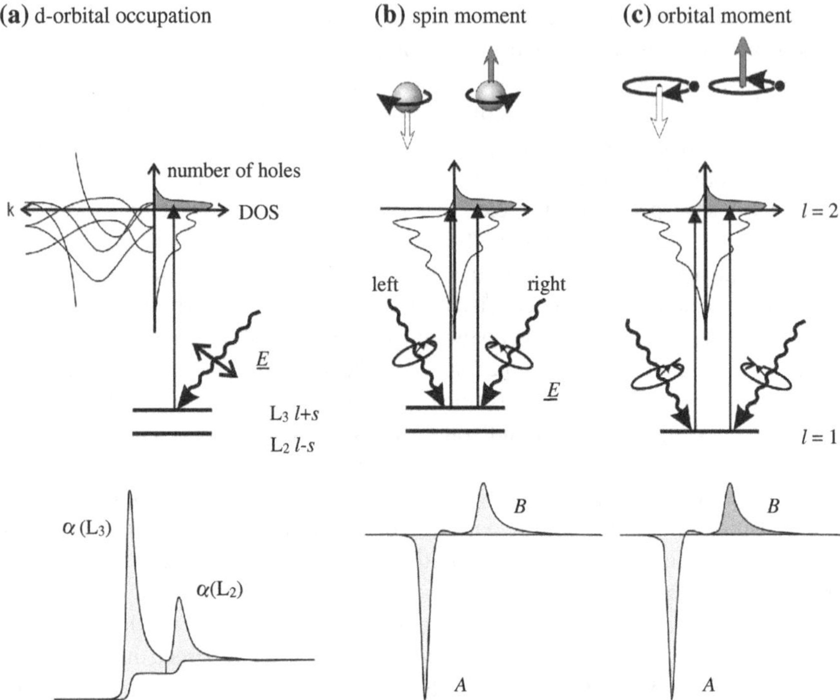

Fig. 5.9 **a** Energy scheme and principle of electronic transitions for conventional L-edge X-ray absorption, **b** and **c** mechanism of XMCD illustrate in the one-electron model approximation. The electrons become excited from the spin-orbit split 2p core shell to empty conduction band states above the Fermi level. The level of absorption is directly related to the number of holes in the d band. From XMCD the spin moment (**b**) and orbital moment (**c**) can be deduced taking into account the dichroic difference intensities A and B, as discussed in the text [34]

For the estimation of the z-component of the spin moment $\langle S_z \rangle$ beside the integrated dichroic difference intensities also the magnetic dipole operator $\langle T_z \rangle$ needs to be taken into account [33]:

$$\langle S_z \rangle = \frac{\int_{j^+} \mathrm{dE}\, \Delta\mu(E) - \frac{l_c+1}{l_c} \int_{j^-} \mathrm{dE}\, \Delta\mu(E)}{\int_{j^+ + j^-} \mathrm{dE}\, (\mu^+(E) + \mu^-(E))} \cdot \frac{2l_c(4l + 2 - n_e)}{l(l+1) - 2 - l_c(l_c+1)} - \frac{l(l+1)[l(l+1) + 2l_c(l_c+1) + 4] - 3(l_c-1)^2(l_c+2)^2}{2l(l+1)[l(l+1) - 2 - l_c(l_c+1)]} \cdot \langle T_z \rangle \quad (5.14)$$

This operator describes the anisotropy field of the spins in a distorted atomic cloud caused either by crystal-field effects or spin-orbit coupling [35]. Based on the general equation above the spin moment can be measured by absorption experiments at the $L_{2,3}$- or $M_{4,5}$-edges using the following expressions:

$$\langle S_z \rangle = \frac{\int_{L_3} dE\,\Delta\mu(E) - 2\int_{L_2} dE\,\Delta\mu(E)}{\int_{L_3+L_2} dE\,(\mu^+(E) + \mu^-(E))} \cdot (10 - n_{3d}) - \frac{7}{2}\langle T_z \rangle, \tag{5.15}$$

$$\langle S_z \rangle = \frac{\int_{M_5} dE\,\Delta\mu(E) - \frac{3}{2}\int_{M_4} dE\,\Delta\mu(E)}{\int_{M_5+M_4} dE\,(\mu^+(E) + \mu^-(E))} \cdot (14 - n_{4f}) - 3\langle T_z \rangle. \tag{5.16}$$

Although the sum rules provide only an estimation of the spin and orbital moment with an accuracy of about 10 % due to the limitation of the underlying single ion model and approximated magnetic dipole operator they offer a sufficient correct comparison of the intrinsic magnetic properties within one material class like thin Fe–Tb alloy films with varying composition.

In order to analyze the magnetic sublattice configuration and the intrinsic magnetic moments in Fe–Tb alloy films and Co/Pt multilayers XMCD measurements were performed at the Fe $L_{2,3}$-edges (719.9 eV, 706.8 eV), Co $L_{2,3}$-edges (793.2 eV, 778.1 eV), and Tb $M_{4,5}$-edges (1276.9 eV, 1241.1 eV) using two different end stations at several beamlines of the Helmholtz-Zentrum Berlin (HZB). The *ALICE diffractometer* installed at the *PM3* beamline allows to obtain XMCD spectra and element specific hysteresis loops with a maximum external field of ±7 kOe in a temperature range from 10 K to 400 K. For investigations in high magnetic fields the *High-field diffractometer* attached to the *UE46-PGM-1* beamline was utilized providing measurements within a maximum external field of ±60 kOe in a temperature range from 4 K to 350 K. The element specific hysteresis loops were taken from intensity measurements of the transmitted circularly polarized light depending on the external magnetic field while the photon energy was fixed to the corresponding absorption edge of Fe (L_3), Co (L_3), and Tb (M_5). The XMCD spectra were recorder in two different ways. At beamline *PM3* the polarization is fixed and the dichroism was deduced from the absorption spectra measured in positive and negative magnetic field. In contrast the *UE46-PGM-1* undulator beamline offers fast polarization changes therefore the spectra could be obtained with constant magnetic field in the *High-field diffractometer.*

5.4.3 Integral Magnetic Measurements Using a Superconducting Quantum Interference Device

Superconducting quantum interference device (SQUID) magnetometer provide the highest sensitivity for magnetic field measurements. The resolution of about 10^{-12} G allows stray field measurements and the estimation of the total magnetic moment of magnetic films with thicknesses in the nanometer range [36]. This kind of magnetometer takes advantage of the quantization of the magnetic flux in a superconducting ring and the Josephson effect, which allows the tunneling of Cooper pairs through a normal conducting barrier. Depending on the operation mode, SQUID magnetometers consist of a superconducting ring interrupted by either one or two Josephson junctions for AC or DC mode. Since the operation principles of such devices are well discussed in the literature [37–39] the following part focuses on the used sample holders with their typical artefacts and the different measurement procedures important for this work.

5.4.3.1 Setup Properties, Sample Holders, and Measurement Artefacts

The integral magnetic properties of the magnetic single layers and exchange coupled heterostructures were obtained using a *MPMS SQUID VSM* [40] from *Quantum Design* with a sensitivity $< 8 \times 10^{-8}$ emu at an external magnetic field of 70 kOe. The standard setup allows magnetic measurements at temperatures between 1.8 and 400 K with an applied field ranging from −70 to 70 kOe.

In order to perform measurements with the applied magnetic field pointing out-of-plane or in-plane with respect to the film surface three different kinds of sample holders can be used. For in-plane measurements the samples become glued onto a half cylindrical quartz rod. Since the total vertical dimension of the pickup coils of the *MPMS SQUID VSM* is smaller than the length of the quartz, the sample holder itself gives almost no contribution to the measurement signal. However, the substrate of the sample as well as the glue produce a diamagnetic background signal as intensively investigated by Christoph Brombacher in 2010 [41]. With regard to this work a careful background subtraction is required especially for magnetic materials with low moments like in Fe–Tb alloy films. Measurements in out-of-plain direction allow the use of two different holders, a brass hollow cylinder and a straw. The brass holder consists of a tube open at one side within two diamagnetic quartz cylinders fixing the sample in between of them. The total length of these cylinders is smaller than the dimension of the pickup coils. This leads to a paramagnetic background caused by the gap between both cylinders. For the measurement of low moment Fe–Tb alloy films such brass sample holders can not be used, since their paramagnetic contribution does not allow a sufficient correct background subtraction. High sensitivity measurements with small background contribution require straw holders. The straws consist of a polymer, which is paramagnetic. Within the straw rod the sample becomes glued in between of two lengthwise folded straws. Due to the missing paramagnetic material in the gap produced by the sample a diamagnetic background signal occurs in the measurement. In general this diamagnetic signal is small compared to the paramagnetic background of the brass holder and can be clearly distinguished from the ferromagnetic signal caused by the magnetic film.

Beside the background signal originated from the sample holders the *MPMS SQUID VSM* shows an artefact in the field range from −1.0 to 1.0 kOe. A sample hysteresis loop of a $Fe_{81}Tb_{19}$(20 nm)/Pt(1 nm)/[Co(0.4 nm)/Pt(0.8 nm)]$_{10}$ heterostructure is given in Fig. 5.10a. The artefact consists of a magnetic soft phase with zero remanence and a saturation moment of about 20 μemu. From XMCD absorption measurements such a soft phase can not be found in the element specific hysteresis loops (Fig. 5.10b) of the material. Therefore the artefact is most likely caused by the measurement system itself. Despite much efforts the origin of this soft

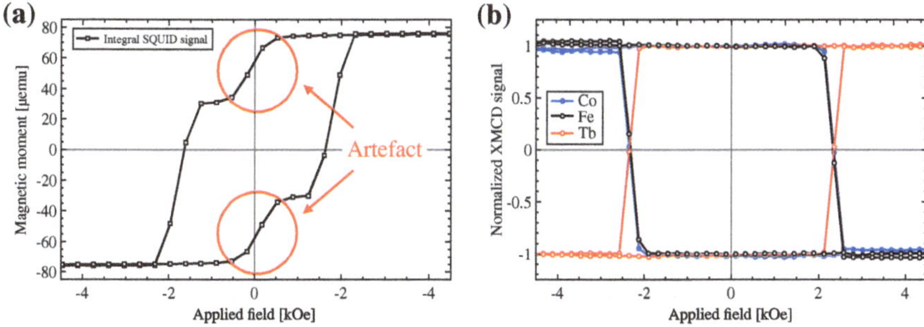

Fig. 5.10 Comparison between the integral **a** and element specific **b** hysteresis loops of a $Fe_{81}Tb_{19}$(20 nm)/Pt(1 nm)/[Co(0.4 nm)/Pt(0.8 nm)]$_{10}$ heterostructure obtained from SQUID magnetometry and XMCD absorption measurements at the Fe and Co L_3-edge and the Tb M_5-edge, respectively. The soft magnetic phase occurs only in the SQUID hysteresis loop and is not caused by the material properties of the heterostructure

magnetic phase is still unknown. With regard to this the SQUID hysteresis loops will be corrected by subtracting the artefact.

5.4.3.2 Measurement Procedures

The saturation magnetization, remanence, coercivity, and in general the shape of the hysteresis loop offer particular information about the magnetic configuration and reversal processes of the magnetic moments in a material. To elucidate this magnetization hysteresis loops were taken with a maximum external field of $\pm$70 kOe in a temperature range from 4 to 400 K. Furthermore, the temperature dependence of the remanence magnetization of Fe–Tb alloy films and Co/Pt multilayers was measured from 4 to 400 K after saturating the samples in an external magnetic field of 70 kOe at 300 K. The superconducting magnet of the SQUID was reset by heating over its transition temperature before the measurement to avoid the influence of a small magnetic field coming from the remanence of the magnet. Providing that the remanence and saturation magnetization are equal the temperature dependence of the net magnetization can be deduced from this experiment. Please note, the remanence and saturation magnetization were always determined in advance using MOKE or SQUID hysteresis loop measurements.

In order to investigate the interfacial exchange coupling, the exchange-bias field, and training effect in magnetic heterostructures consisting of Fe–Tb alloy films and Co/Pt heterostructures a particular measuring procedure was used. All samples were saturated at 300 K in an applied field of 70 kOe and afterwards cooled down in a certain cooling field to the desired measuring temperature. At this temperature hysteresis measurements were performed with different maximum magnetic fields and number of cycles. This procedure was repeated for each changing parameter like

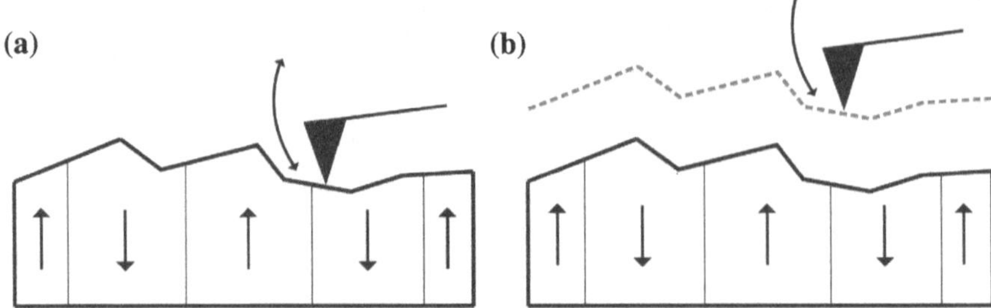

Fig. 5.11 Schematics of the *Lift Mode* **a** the first line scan obtains a height profile of the surface and **b** the second line scan measures the magnetic signal by following the topography in a distance between 20 and 30 nm from the surface [41]

temperature, cooling field, or maximum applied field, which guaranties the same starting point and history for all measurements.

5.4.4 Stray Field Imaging with Magnetic Force Microscopy

As mentioned above the magnetic force microscopy (MFM) is a special type of the AFM, which allows the imaging of magnetic stray fields due to dipolar interactions between a magnetic sample and a magnetized tip. In particular the gradient of the magnetic stray field produce a force on the magnetic tip changing the eigenfrequency of the cantilever. Since literature [15, 42–45] provides a variety of background information concerning the physical mechanisms and technical implementation of MFM the following part will exclusively focus on the measurement setups and procedures used in this work.

The domain configuration of magnetic thin films and patterning in the demagnetized state were measured with the same AFM setup as previously discussed. In order to separate the magnetic signal from Van der Waals and other forces producing a topographical contrast the setup can be run in the so called *Lift Mode*. After the topography is imaged in the first line scan, the tip becomes retracted to a distance between 20 and 30 nm from the surface, and a second line scan is performed with a constant tip-sample distance based on the topography information obtained previously. The principle of the first and second line scan is presented in Fig. 5.11. The lateral resolution depending on the tip-sample distance, tip shape and damping constant of the cantilever is limited to about 20 nm in this setup. For the measurements Co coated MFM tips from *Team Nanotec GmbH* with a radius better than 40 nm and a damping constant of 3 N/m were used.

Additionally, high-resolution in-field MFM investigations of Fe–Tb nanodot arrays were carried out in a HR-MFM system from *NanoScan Ltd.* with a lateral resolution of 10 nm [46]. For high-resolution the system operates in vacuum with a base pressure better than 10^{-5} mbar and uses ultra low moment tips from *Team*

Nanotec GmbH possessing a radius better than 25 nm and a damping constant of 0.7 N/m. The microscope allows measurements within perpendicular magnetic fields up to 5 kOe. Instead of *Tapping Mode* the *Constant Height Mode* is utilized with an applied electrical bias potential to compensate electrostatic interactions. In this mode the cantilever performs line scans with a constant average distance between the tip and the sample surface measuring the shift of the eigenfrequency due to the dipolar interactions. Because of the larger distance between tip and surface as compared to the *Tapping Mode* the influence by the topography is almost negligible providing pure magnetic contrast.

References

1. Bestec GmbH, Sputtering systems. http://www.bestec.de. Accessed 01 Aug 2012
2. S.Y. Chou, P.R. Krauss, P.J. Renstrom, Science **272**, 85 (1996)
3. L.J. Guo, J. Phys. D: Appl. Phys. **37**, R123 (2004)
4. L.J. Guo, Adv. Mater. **19**, 495 (2007)
5. M. Altissimo, Biomicrofluidics **4**, 026503 (2010)
6. H. Kiessig, Ann. Phys. **10**, 769 (1931)
7. K. Oura, V.G. Lifshits, A.A. Saranin, A.V. Zotov, M. Katayama, *Surface Science: An Introduction* (Springer, Berlin, 2003). ISBN 978-3-540-00545-2
8. M. Mayer, SIMNRA user's guide (1997)
9. M. Birkholz, *Thin Film Analysis by X-Ray Scattering* (WILEY-VCH Verlag GmbH & Co, KGaA, Weinheim, 2006). ISBN 3-527-31052-5
10. B.Warren, X-ray Diffraction (1990). ISBN 978-0-486-66317-3
11. R. Erni, M. Rossell, C. Kisielowski, U. Dahmen, Phys. Rev. Lett. **102**, 096101 (2009)
12. M.T. Otten, P.M. Mul, M.J. de Jong, Microsc. Microanal. Microstruct. **3**, 83 (1992)
13. FEI Company, Nova NanoSEM. http://www.fei.com/uploadedfiles/documents/case_studies/mintek_07_2008_cs.pdf. Accessed 09 Sept 2012
14. Y. Seo, W. Jhe, Rep. Prog. Phys. **71**, 016101 (2007)
15. U. Hartmann, Annu. Rev. Mater. Sci. **29**, 53 (1999)
16. J. Kerr, Phil. Mag. **3**, 321 (1877)
17. Z.Q. Qiu, S.D. Bader, Rev. Sci. Instrum. **71**, 1243 (2000)
18. P. Weinberger, Phil. Mag. **88**, 897 (2008)
19. T. Ulbrich, Aufbau eines MOKE-Systems zur Untersuchung magnetischer Nanostrukturen, Diploma Thesis, Universität Konstanz (2003)
20. F. Springer, Magnetic and structural properties of granular CoCrPt films, Diploma Thesis, Universität Konstanz (2007)
21. G. Schütz, W. Wagner, W. Wilhelm, P. Kienle, R. Zeller, R. Frahm, G. Materlik, Phys. Rev. Lett. **58**, 737 (1987)
22. P. Zeeman, Nobel Lecture (1902)
23. R. Wu, D. Wang, A. Freeman, Phys. Rev. Lett. **71**, 3581 (1993)
24. G.Y. Guo, H. Ebert, W.M. Temmerman, P.J. Durham, Phys. Rev. B **50**, 3861 (1994)
25. R. Wu, D. Wang, A.J. Freeman, J. Magn. Magn. Mater. **132**, 103 (1994)
26. B. Thole, G. Van der Laan, G. Sawatzky, Phys. Rev. Lett. **55**, 2086 (1985)
27. G. van der Laan, B. Thole, G. Sawatzky, J. Goedkoop, J. Fuggle, J.-M. Esteva, R. Karnatak, J. Remeika, H. Dabkowska, Phys. Rev. B **34**, 6529 (1986)
28. J.B. Goedkoop, J.C. Fuggle, B.T. Thole, G. Van der Laan, G.A. Sawatzky, J. Appl. Phys. **64**, 5595 (1988)
29. M. Sacchi, R.J.H. Kappert, J.C. Fuggle, E.E. Marinero, Appl. Phys. Lett. **59**, 872 (1991)

30. M. Sacchi, O. Sakho, G. Rossi, Phys. Rev. B **43**, 1276 (1991)
31. H. Ebert, J. Stöhr, S. Parkin, M. Samant, A. Nilsson, Phys. Rev. B **53**, 16067 (1996)
32. B. Thole, P. Carra, F. Sette, G. Van der Laan, Phys. Rev. Lett. **68**, 1943 (1992)
33. P. Carra, B. Thole, M. Altarelli, X. Wang, Phys. Rev. Lett. **70**, 694 (1993)
34. J. Stöhr, J. Magn. Magn. Mater. **200**, 470 (1999)
35. J. Kanamori, *Anisotropy and Magnetostriction of Ferromagnetic and Antiferromagnetic Materials, Magnetism* (Academic, New York, 1963)
36. C. Enss, S. Hunklinger, *Tieftemperaturphysik* (Springer, Berlin, 2000). ISBN 3-540-67674-0
37. R. Kleiner, Proc. IEEE **92**, 1534 (2004)
38. R.L. Fagaly, Rev. Sci. Instrum. **77**, 101101 (2006)
39. A. Edelstein, J. Phys.: Condens. Matter **19**, 165217 (2007)
40. Quantum Design, MPMS®, SQUID VSM Brochure (2010)
41. C. Brombacher, Rapid thermal annealing of FePt and FePt/Cu thin films, Ph.D. Thesis, TU Chemnitz (2010)
42. Y. Martin, H.K. Wickramasinghe, Appl. Phys. Lett. **50**, 1455 (1987)
43. A. Wadas, P. Grütter, Phys. Rev. B **39**, 12013 (1989)
44. D. Rugar, H.J. Mamin, P. Guethner, S.E. Lambert, J.E. Stern, I. McFadyen, T. Yogi, J. Appl. Phys. **68**, 1169 (1990)
45. A. de Lozanne, Microsc. Res. Tech. **69**, 550 (2006)
46. NanoScan Ltd., HR-MFM. http://www.nanoscan.ch/products/hr-mfm.php. Accessed 24 Sept 2012

Chapter 6
Magnetic Order in Thin $Fe_{100-x}Tb_x$ Films: A Temperature and Stoichiometry Dependent Study

The amorphous structure of thin FI RE-TM alloy films gives rise for manifold magnetic properties and peculiar spin structures (i.e., sperimagnetism) as discussed in Chap. 2. Despite the enormous number of scientific studies performed in this field of magnetic materials, a detailed investigation of the intrinsic magnetic properties and magnetization reversal processes as a function of the temperature and stoichiometry is missing. In particular the magnetic configuration and reversal mechanism in the vicinity of the compensation point is nearly unexplored. With regard to this continuous amorphous $Fe_{100-x}Tb_x$ films with thicknesses of about 20 nm were investigated in a composition range $14 \leq x \leq 35$ at.% Tb with varying temperatures from 4 to 400 K. The sample fabrication was realized by magnetron co-sputtering using standard Si(100) wafers with a 100-nm-thick thermally oxidized SiO_2 layer as substrates. The Fe–Tb alloy films are embedded in Pt seed and capping layers with a nominal thickness of 5 and 3 nm, respectively, which serve as a oxidation barrier. The main results concerning the structural and magnetic properties will be discussed.

6.1 Morphology and Structural Properties

Transmission electron microscopy investigations of the Fe–Tb films reveal an amorphous structure without any hint of micro crystallinity independent from the stoichiometry of the alloy. As an example Fig. 6.1 shows a lateral image in (a) and cross-section images in (c) and (d) as well as a diffraction pattern in (b) of a Pt(3 nm)/$Fe_{75.5}Tb_{24.5}$(20 nm)/Pt(5 nm) layer stack. Since no contrast originates from the amorphous Fe–Tb layer the lateral TEM image reveals the grain structure of the Pt seed and capping layers. The size of the visible Pt grains ranges from 5 to 10 nm, which agrees to results published in literature [1]. The diffraction pattern obtained in lateral geometry exhibits a multi-ring structure according to different crystalline orientations. The (111) diffraction ring is the strongest in the pattern indicating that more Pt grains are oriented in this direction. This is consistent with results reported in literature [2].

C. Schubert, *Magnetic Order and Coupling Phenomena*, Springer Theses,
DOI: 10.1007/978-3-319-07106-0_6,

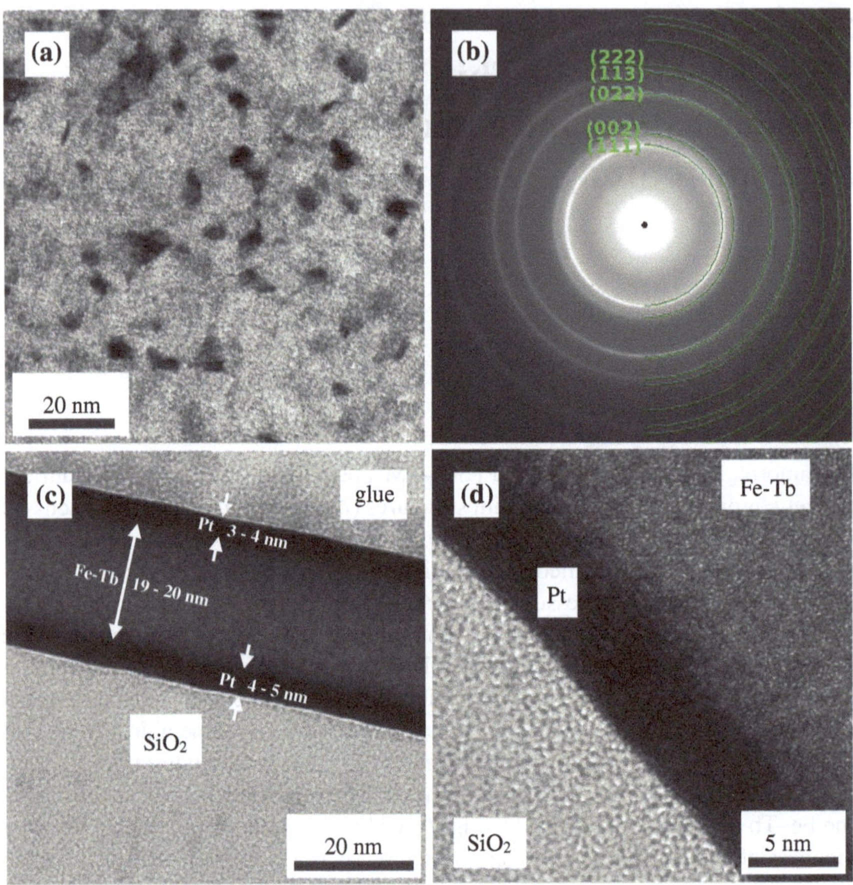

Fig. 6.1 Structural properties and morphology of amorphous Fe–Tb alloy films investigated by TEM imaging. The results on Pt(3 nm)/$Fe_{75.5}Tb_{24.5}$(20 nm)/Pt(5 nm)/substrate are given as an example with **a** lateral image, **b** diffraction pattern indicating polycrystalline Pt layers with (111) texture, and **c**–**d** cross-section images in two different magnifications revealing contineous Pt seed and capping layers and somewhat intermixed interfaces to the amorphous Fe–Tb layer [3]

From the cross-section TEM images presented in Fig. 6.1c, d a continuous layer structure can be found. The film thicknesses agree well to the nominal layer thicknesses. The Pt seed and capping layers form closed films and provide a good oxidation barrier against the atmosphere and the substrate for the entire Fe–Tb layer. The higher magnification in (d) reveals the granular structure of the Pt seed layer with grains of different crystalline orientation as also obtained from the diffraction pattern in (b). Contrary to this, the Fe–Tb layer exhibits neither a crystalline nor a granular structure. Furthermore, a slight roughness can be found at the interface due to inter diffusion between Pt and the Fe–Tb alloy during deposition caused by the high energetic sputter atoms.

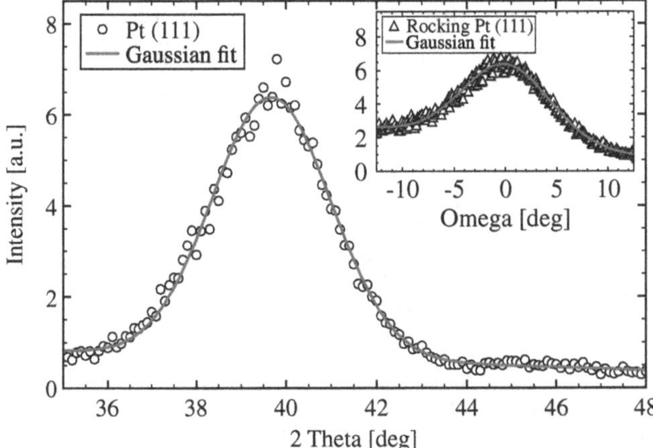

Fig. 6.2 X-ray diffraction pattern of amorphous Fe–Tb films with Pt seed and capping layers. The results on Pt(3 nm)/$Fe_{75.5}Tb_{24.5}$(20 nm)/Pt(5 nm)/substrate serve as an example. The XRD pattern reveals one peak referring to the (111) Bragg reflection of the Pt layers. The inset shows the broad rocking curve of the Pt (111) peak

The (111) texture of the Pt layers is also found by conventional $\theta/2\theta$ XRD measurements. The XRD pattern outlined in Fig. 6.2 reveals a single peak at about $2\theta = 39.68°$, which is attributed to the Bragg reflection of fcc Pt grains oriented in [111] direction. Grains with other orientations yield no reflection in the pattern due to their smaller amount and grain size. According to the results of the TEM investigation no reflexes originating from the Fe–Tb layer are observed. The rocking curve related to the Pt (111) peak shown in the inset of Fig. 6.2 possesses a relative broad distribution. Following the Scherrer equations 5.7 and 5.8 for the cross-sectional and lateral coherence length the FWHM of the peak in the $\theta/2\theta$ scan and the rocking curve allows the estimation of the average grain size in out-of-plane and in-plane direction to 2.4 and 2.2 nm, respectively. Compared to the size of the grains observed in the lateral TEM image in Fig. 6.1a the grains with (111) orientation build only a portion of the grain structure visible by absorption contrast. The rest might consist of Pt grains with other orientations, which are X-ray amorphous and provide no contribution to the XRD pattern due to their smaller size. Furthermore, the lateral grain size was estimated as minimum value assuming that the peak broadening of the rocking curve is exclusively caused by the coherence length. The broadening of the peak can also result from a slight tilt of the Pt (111) grains.

As the magnetic properties of the amorphous Fe–Tb alloy films are strongly dependent on the growth conditions, the underlying seed layer provides a strong influence. However, the exact dependency requires a detailed investigation, which is not the subject of this work. In order to exclude a growth induced variance of the magnetic properties of the Fe–Tb films, the same growth conditions were applied for the various sample series.

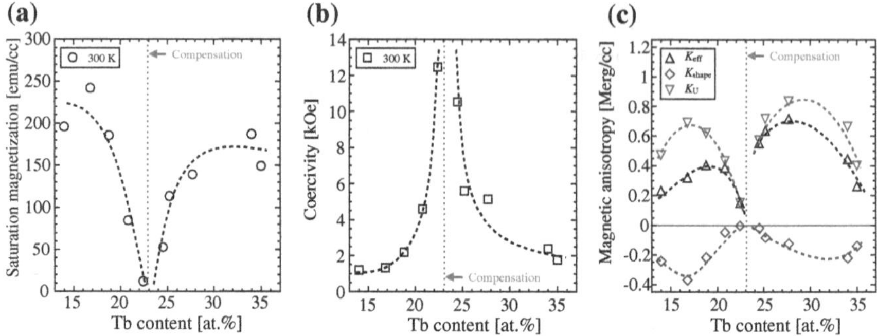

Fig. 6.3 Magnetic properties of 20-nm-thick amorphous Fe–Tb alloy films at RT **a** net saturation magnetization, **b** coercivity, and **c** different magnetic anisotropies as a function of the Tb content. The values were extracted from magnetic hysteresis loops measured in out-of-plane geometry at RT using SQUID magnetometry. Please note, the error values of the measurements are in the range of the symbol size. The dashed curves serve as a guid to the eye and the three vertical dashed lines indicate the RT compensation composition at about 23 at.% Tb

6.2 Sperimagnetism and its Properties at Room Temperature

The short range order expected in the amorphous structure of the Fe–Tb alloy films determines the sperimagnetic spin configuration based on the competition between exchange coupling and single ion anisotropy as described in Chap. 2. Since the amount of the voluminous Tb atoms in the alloy changes the growth conditions for the thin film, the stoichiometry can strongly influence the short range order in the amorphous alloy and with it its magnetic properties like net saturation magnetization, coercivity, and magnetic anisotropy.

6.2.1 Net Saturation Magnetization, Coercivity, and Magnetic Anisotropy with Respect to the Tb Content of the Fe–Tb Alloy Film

In general, the net saturation magnetization of a FI alloy film with collinear magnetic sublattices is determined by the amount of atoms and their magnetic moment of each sublattice. Non-collinear spin structures possess an additional influence due to the average opening angle of the fanning cone structure, which reduces the net saturation magnetization. This has to be taken into account for the discussion of the magnetic properties of the amorphous Fe–Tb alloy system.

Figure 6.3 shows in (a) the net saturation magnetization, in (b) the coercivity, and in (c) the effective, shape, and uniaxial magnetic anisotropy of 20-nm-thick amorphous $Fe_{100-x}Tb_x$ alloy films as a function of the Tb content at RT. Starting from a small amount of Tb, the net saturation magnetization becomes reduced to zero with increasing Tb content towards the RT compensation point at about 23 at.% Tb where the antiparallel oriented Fe and Tb sublattices compensate each other. Above the compensation point the dominant sublattice magnetization changes from Fe to Tb. And with further increasing amount of Tb the net magnetization increases again. The RT compensation point at 23 at.% Tb is in good agreement to the literature [4]. However, the exact value depends strongly on the growth conditions like Ar sputter pressure, temperature, deposition rate, and the amount of oxygen in the residual atmosphere due to the influence of the short range order and the fanning cone structure.

Since net saturation magnetization and coercivity follow a reciprocal relation (see Eq. 2.2) the coercivity increases strongly towards the compensation point as it is observed in Fig. 6.3b. The magnetic anisotropy, shown in Fig. 6.3c, exhibits also a stoichiometry dependency. The shape anisotropy of the amorphous Fe–Tb film decreases equally to the net saturation magnetization in the vicinity of the compensation point. The effective anisotropy, which was calculated as lower estimate from the net saturation magnetization and the coercivity following Eq. 2.2, tend to become reduced near the compensation point. In the literature both cases minimum [5] and maximum [6] magnetic anisotropy at the compensation point are reported. However, an accurate value can not be given, since near the compensation point the in-plane and out-of-plane saturation fields exceed the maximum possible external magnetic field of 70 kOe of the SQUID magnetometer. Furthermore, the reversal mechanism changes making it hard to define an anisotropy as will be discussed later.

Towards lower and higher Tb contents the effective and uniaxial magnetic anisotropy of the Fe–Tb alloy films become reduced again. Nevertheless, even films with 14 and 35 at.% Tb exhibit an out-of-plane anisotropy, which is in contrast to the results of Y. Mimura et al. [7] where only $Fe_{100-x}Tb_x$ films with 15 at.% $\leq x \leq$ 29 at.% are reported as out-of-plane. A reason for this difference might be the film thickness. The films reported by Y. Mimura et al. with thicknesses of about 140 nm were substantially thicker compared to the 20-nm-thick Fe–Tb films of this investigation. Since the anisotropy in the short range order induced by the growth process causes the magnetic anisotropy, [8–11] an increasing film thickness, which allows the amorphous system to relax more and more into the isotropic state, reduces the magnetic anisotropy. Therefore, thinner films are expected to exhibit a higher intrinsic magnetic anisotropy perpendicular to the film plane. As the competing shape anisotropy increases towards lower and higher amounts of Tb according to the increasing net saturation magnetization, thicker Fe–Tb films with their smaller intrinsic anisotropy exhibit an effective PMA in a smaller composition region around the compensation point compared to thinner films. Please note that the increase of the intrinsic anisotropy with decreasing film thickness is only valid in a limited thickness range. Preliminary investigations on Fe–Tb alloy films with varying thicknesses (not shown) revealed that the effective PMA decreases strongly if the film thickness becomes smaller than 20 nm.

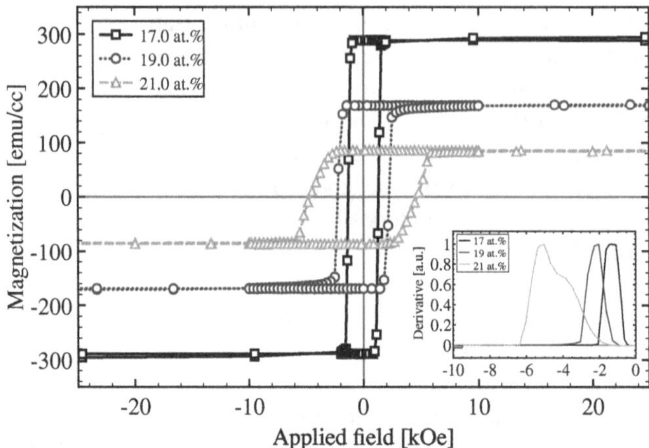

Fig. 6.4 Magnetic hysteresis loops of 20-nm-thick $Fe_{100-x}Tb_x$ films with a Tb content x of 17, 19, and 21 at.% obtained from SQUID magnetometry measurements at RT. The inset shows the derivative of the magnetization curves in the negative field range providing a measure of the branch width and information about the reversal process

6.2.2 Magnetization Reversal in Fe–Tb Films Dominated by the Fe Sublattice

The magnetization reversal of amorphous Fe–Tb films with a Tb content below 23 at.% is determined by the dominant Fe sublattice interacting with the external magnetic field as mentioned in the previous paragraph. An exemplary overview of RT magnetization hysteresis loops of 20-nm-thick $Fe_{100-x}Tb_x$ films with a Tb content x of 17, 19, and 21 at.% is given in Fig. 6.4. The inset shows the derivative of the magnetization curves in the negative field range providing a measure of the branch width and information about the reversal process. In general Fe–Tb alloy films with small amounts of Tb like 17 at.% exhibit high net saturation magnetization, low coercivity, and a rather sharp reversal branch. This indicates a magnetization reversal driven by domain nucleation followed by wall propagation. This behavior is found commonly for thin exchange coupled magnetic films with PMA [12, 13]. In comparison to the cross-section TEM images presented in Sect. 6.1, which reveal continuous layers without a granular structure, the amorphous Fe–Tb films can be assumed as fully exchange coupled. Due to the amorphous structure and missing grain boundaries the magnetization reversal process is nearly unaffected by domain wall pinning and the domain nucleation appears most likely at macroscopic defects and the sample border. Therefore, the wall propagation field is smaller compared to the nucleation field and the reversal takes place instantaneously as soon as the external applied magnetic field reaches the nucleation field.

With increasing Tb content the net saturation magnetization decreases towards the compensation point. In consequence the driving force of the external field due to the Zeeman energy becomes reduced and the coercivity increases. Furthermore, the width of the switching field distribution (SFD) is broadened significantly for Fe–Tb films with reducing net saturation magnetization below 100 emu/cc as shown for a film stoichometry of 21 at.% Tb in Fig. 6.4 by the hysteresis loop and the corresponding derivative in the inset. This indicates a change in the magnetization reversal mechanism. Due to the amorphous structure of the films, similar low defect densities are to be expected independently from the Tb content. Therefore, the amount of pinning sites in the films is most likely the same and additional domain wall pinning effects changing the reversal process in the vicinity of the compensation point can be excluded. Especially stray field induced domain wall pinning at non-magnetic defects becomes reduced with decreasing net saturation magnetization.

The reduction of the strength of the stray field according to the magnetization towards higher Tb content leads to different domain stability conditions, which allows larger domain sizes near the compensation, as observed by B. Lanchava and H. Hoffmann [14]. Accompanied with the size of the domains its shape changes from stripe to bubble like, which can explain the different reversal processes. The stripe domain reversal in this kind of amorphous system appears most likely by stripe length expansion via wall motion after the nucleation of a variety of small bubble domains [15]. Magnetization switching in this manner provides a rather sharp reversal branch with narrow SFD as found in the $Fe_{100-x}Tb_x$ alloy films with small Tb content ($x \leq 19$ at.%). Contrary to this the reversal of $Fe_{100-x}Tb_x$ films with low net saturation magnetization (21 at.% $\leq x <$ 23 at.%) behaves differently. After nucleation of a variety of small bubbles no stripe domains occur [14]. With increasing external magnetic field several bubbles grow and merge, which leads to the complete magnetization reversal. The dynamic of both processes, the nucleation and growth of the bubble domains during the reversal can be deduced from the derivative of the magnetization hysteresis loop of the Fe–Tb film with 21 at.% Tb in the inset of Fig. 6.4. The peak in the derivative reveals a broad shoulder most likely attributed to the bubble nucleation and expansion. The peak maximum itself represents the irreversible part of the magnetization switching when the bubble domains start to merge. This particular reversal mechanism and the small driving force of the external magnetic field due to the small net saturation magnetization leads to the observed broad SFD in the vicinity of the compensation point.

6.2.3 Magnetization Reversal in Fe–Tb Films Dominated by the Tb Sublattice

Above the RT compensation composition of 23 at.% Tb the interaction between the external magnetic field and the net saturation magnetization of the Fe–Tb films becomes dominated by the Tb sublattice. As the magnetic moments are non-collinear oriented in the Tb sublattice, the reversal process becomes strongly determined by the particular structure of the fanning cone, which was introduced in Sect. 2.2.

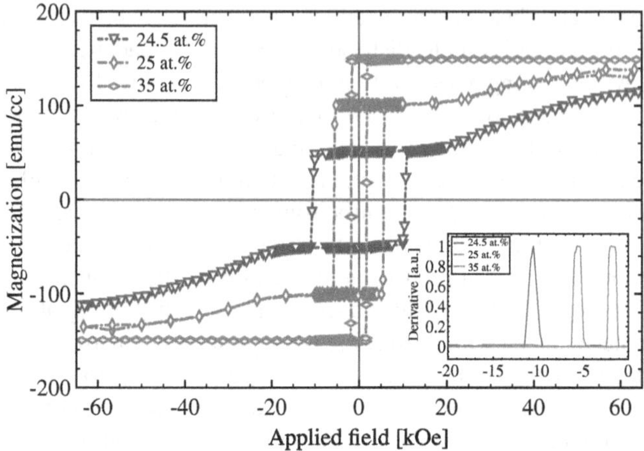

Fig. 6.5 Magnetic hysteresis loops of 20-nm-thick $Fe_{100-x}Tb_x$ films with a Tb content x of 24.5, 25, and 35 at.% obtained from out-of-plane SQUID magnetometry measurements at RT. The inset shows the derivative of the magnetization curves in the negative field range providing a measure of the branch width and information about the reversal process

An exemplary overview of the magnetization hysteresis loops of Tb dominated $Fe_{100-x}Tb_x$ films is given in Fig. 6.5 with compositions of 24.5, 25, and 35 at.% Tb. The inset contained in the graph reveals the derivative of the magnetization curves. As described in Sect. 6.2.1 the saturation magnetization becomes enhanced with increasing Tb content starting from the compensation point. Beside the amount of Tb in the alloy the average orientational distribution of the Tb moments has also an impact on the magnetization. Contrary to Fe–Tb films with dominant Fe sublattice magnetization the main reversal process is not changing with Tb concentration. Even in the vicinity of the compensation point the reversal branch remains more or less unchanged according to the derivative of the magnetization. The reversal is most likely driven by domain nucleation followed by wall propagation. Only the nucleation field becomes larger with decreasing net saturation magnetization due to the reduced driving force of the external magnetic field. However, an additional feature in the high magnetic field regime is observed near compensation. Above a field of 20 kOe the magnetization of Fe–Tb films with less than 25 at.% Tb increases further with respect to the remanence value. The enhancement reveals full reversibility. Following the work of J. M. D. Coey [16] such a behavior originates from the interaction of the fanning cone structure with the external field. With increasing magnetic field the opening angle of the fanning cone decreases, which results in an enlargement of the net magnetization. This process shows a reversible characteristic as at zero field the fanning cone structure relaxes in its equilibrium state.

The influence of the external field on the fanning cone depends strongly on its equilibrium structure, strictly speaking, the average width of the orientational moment distribution in zero field, which is determined by the single ion anisotropy and exchange coupling strength (see Sect. 2.2). In consequence this distribution is related

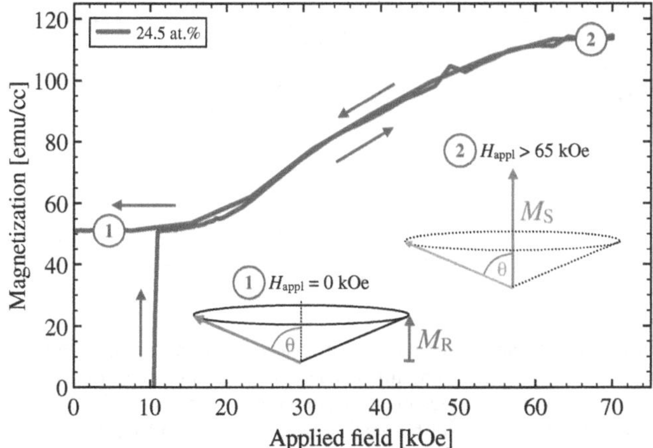

Fig. 6.6 Magnetic hysteresis loops of a 20-nm-thick $Fe_{75.5}Tb_{24.5}$ film in the positive field range obtained from out-of-plane SQUID magnetometry measurements at RT. The schematic drawing denoted with number 1 and 2 outline the magnetic configuration of the dominant Tb sublattice moments in equilibrium zero field state and in saturation at magnetic fields above 65 kOe. The average opening angle of the orientational Tb moment distribution (*fanning cone*) is approximated in a macro-spin simplification and completely parallel aligned magnetic moments are assumed in saturation

to the structure of the short range order induced by the growth conditions and with it to the Tb content of the alloy. By comparing the hysteresis loops of $Fe_{100-x}Tb_x$ films with 24.5 and 25 at.% Tb in Fig. 6.5 it turns out that towards smaller Tb contents and net saturation magnetizations in the vicinity of the compensation point the relative change of the magnetization in the high field regime increases.

Assuming a complete collinear orientation of the Tb moments above 65 kOe and considering a macro-spin simplification as outlined by the schematics in Fig. 6.6 a minimum average opening angle θ of the fanning cone in zero field can be estimated according to the following equation:

$$\theta = \arccos \frac{M_R}{M_S}, \tag{6.1}$$

where M_R and M_S denotes the remanence and saturation magnetization resulting from the projected out-of-plane net moment, which dependents on the average opening angle of the fanning cone structure with regard to the external applied magnetic field H_{appl}. Based on this assumptions and the magnetization hysteresis loops in Fig. 6.5 the average fanning cone opening angle of the Tb moments of thin Fe–Tb alloy films with 24.5 and 25.0 at.% Tb results to about $60° \pm 5°$ and $40° \pm 5°$, respectively. The significant change in the fanning cone structure towards the RT compensation point originates most likely from the reduced net saturation magnetization and the weak stray field interaction with the magnetic moment and single

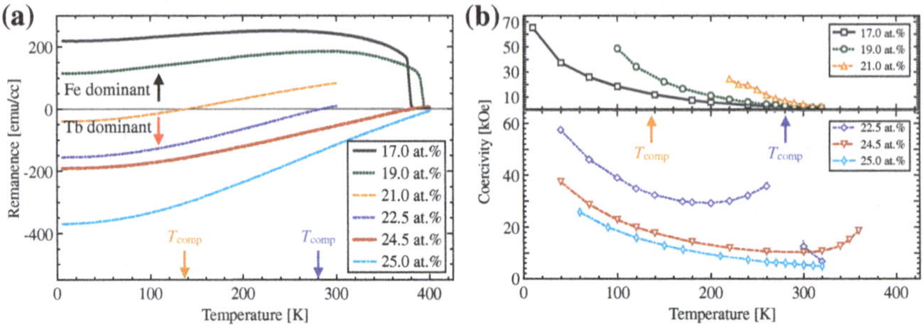

Fig. 6.7 **a** Out-of-plane remanent net magnetization as a function of temperature and composition obtained from 20-nm-thick $Fe_{100-x}Tb_x$ films. The remanence magnetization was measured from 4 to 400 K after saturating the samples in an external magnetic field of 70 kOe at RT. The magnet of the *MPMS SQUID VSM* was reset by heating over its superconducting transition temperature before the measurement to avoid the influence of a small magnetic field coming from the remanence of the magnet. The positive and negative region of the net magnetization refers to a dominant Fe and Tb sublattice, respectively. **b** Corresponding coercivity of the Fe–Tb alloy films obtained from hysteresis loop measurements at different temperatures. The *lines* act as a guid to the eye and the arrows indicate the observed compensation temperatures. Furthermore, the progress of the coercivity is separated in the diagram for Fe (*upper part*) and Tb (*lower part*) dominant alloy films

ion anisotropy of the Tb atoms during film growth. This might allow the observed broadening of the orientational distribution of the Tb moments.

6.3 The Influence of Temperature on the Sperimagnetic Configuration

Beside stoichometry, net saturation magnetization, and growth conditions, the temperature has also a significant impact on the structure of the magnetic fanning cone of the Fe–Tb alloy films. As the fanning cone structure determines the total magnetization of the Tb sublattice, changes in temperature also influence the overall magnetic properties of the Fe–Tb alloy films.

The temperature and stoichiometry dependence of the out-of-plane remanent net magnetization and coercivity of 20-nm-thick $Fe_{100-x}Tb_x$ films are presented in Fig. 6.7a, b, respectively. According to the temperature dependence of the net magnetization and the Tb content in the range from 14 to 35 at.%, the alloy system can be classified with respect to the dominant magnetic moment influencing the existence of a magnetization compensation point in the temperature range up to the Curie temperature, which is about 400 K [7]. $Fe_{100-x}Tb_x$ alloy films with $x \leq 20$ at.% reveal no magnetization compensation point. The net magnetization is dominated by the Fe sublattice. Towards lower temperatures the net saturation magnetization decreases, which is attributed to an increase of the Tb sublattice magnetization. In a composition range of 20 at.% $< x \leq 24$ at.% the Fe–Tb alloys show a change of the dominant moment from Tb to Fe with increasing temperature. This is indicated by the net magnetization passing zero at the magnetization compensation temperature $T_{\rm comp}$. For alloys with a Tb content of $x > 24$ at.%, the Tb sublattice becomes dominant in the whole temperature range and the net magnetization increases with decreasing temperature due to the gain in magnetic moment of the Tb sublattice. This agrees well to results presented in the literature [7].

In general, the Tb atoms of the amorphous Fe–Tb alloy films possess a magnetic moment of almost 9 $\mu_{\rm B}$/atom, the value of a free ionized Tb atom [17]. Due to the strongly localized 4f electrons, the magnetic exchange between Fe and Tb atoms takes place indirectly via 3d – 5d – 4f hybridization [18]. Therefore, the magnetic 4f moments of the Tb atoms remain nearly unaffected by the chemical order and electronic structure of the adjacent Fe atoms. This allows the high magnetic moment of the Tb atoms in the alloy. As described in Sect. 2.2 the magnetic configuration of the Fe–Tb alloy system and in particular the structure of the Tb sublattice arise from the interaction between the single ion anisotropy of the Tb atoms and the magnetic exchange coupling leading to the orientational distribution of magnetic moments referred to as fanning cone structure. With regard to this, one can assume that the total magnetic moment per Tb atom stays more or less constant with temperature and the fanning cone structure determines mainly the temperature dependence of the out-of-plane net saturation magnetization presented in Fig. 6.7a. Although the magnetic moment per Fe atom varies also with temperature, a stronger influence on the net magnetization is given by the change of the average opening angle of the orientational distribution of the Tb moments. With decreasing temperature the exchange coupling increases [19] and according to Eq. 2.1 the fanning cone becomes narrower leading to an increase of the total value of the Tb sublattice magnetization in out-of-plane direction. Depending on the stoichometry of the Fe–Tb alloy film the influence of the fanning cone structure is more or less present, which causes the different temperature characteristics observed in the net saturation magnetization.

The temperature dependence of the coercivity for Fe–Tb films with different compositions, as shown in Fig. 6.7b, reveal similarities independent of the particular Tb content. The coercivity is mainly influenced by the net saturation magnetization and the effective magnetic anisotropy. Both contributions manifest themselves in a different manner. In the vicinity of the compensation point the coercivity becomes strongly enhanced, which originates from the small net saturation magnetization and the reduced driving force of the external magnetic field according to Eq. 2.2. This kind of coercivity enhancement can be seen in the temperature curves for Fe–Tb films with 22.5 and 24.5 at.% Tb in the vicinity of the compensation points at 280 and 380 K, respectively. This particular enhancement becomes superimposed by the influence of the magnetic anisotropy on the coercivity. Independent from the existence of a compensation point, the coercivity increases towards lower temperatures due to a strong gain in the effective PMA. This enhancement is a characteristic of the FI amorphous alloy system and hardly influenced by the stoichiometry. In consequence Fe–Tb films with 22.5, 24.5, and 25 at.% Tb, which possess no compensation point in the low temperature regime, show a similar temperature dependence in coercivity (see Fig. 6.7b). Contrary to this, the Fe dominated films with low

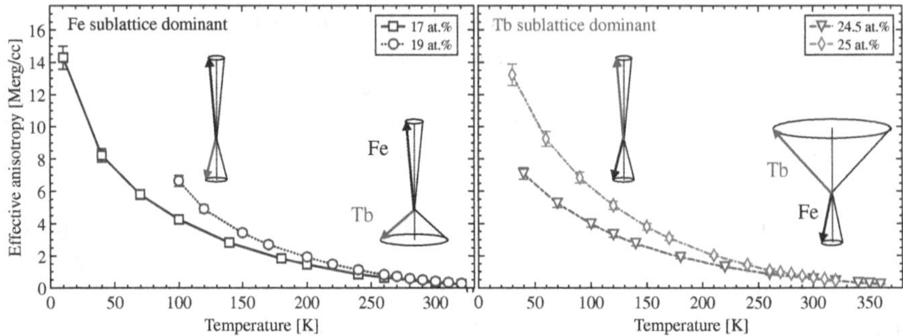

Fig. 6.8 Effective magnetic anisotropy depending on the temperature of 20-nm-thick amorphous $Fe_{100-x}Tb_x$ films with varying Tb content x. The alloy films with 17 and 19 at.% Tb possess a net magnetization dominated by the Fe sublattice moments in the whole temperature range up to the Curie temperature (*left graph*), films containing Tb amounts of 24.5 and 25 at.% reveal a Tb dominant sublattice (*right graph*). Please note, error values greater than the size of the symbols are indicated with *bars*

temperature compensation points (e.g., $T_{comp} = 140\,K$ for $x = 21$ at.%) exhibit a coercivity enhancement due to the increasing magnetic anisotropy superimposed by the influence of the compensation point, which leads to very different characteristics. Generally the magnitude of the coercivity at a certain temperature for the amorphous Fe–Tb alloy films depends on the corresponding net magnetization, the value of the compensation temperature, and the magnetic anisotropy.

The minimum effective magnetic anisotropy of thin amorphous $Fe_{100-x}Tb_x$ alloy films estimated from coercivity and net saturation magnetization according to Eq. 2.2 is shown in Fig. 6.8 as a function of the stoichiometry and temperature. As already mentioned, the anisotropy increases towards lower temperatures independently from the stoichiometry and dominant magnetic sublattice. At low temperatures anisotropy values above 10 Merg/cc are reached. The schematics in the graphs illustrate the fanning cone structure of the Fe and Tb sublattices in macro-spin approximation with orientational moment distribution. Towards lower temperatures the distribution of the Tb moments becomes smaller due to the increasing magnetic exchange coupling, [19] which consequently leads to the observed increase in PMA similar to the Tb sublattice magnetization discussed above. The fanning cone of the Fe sublattice is assumed as very small and not varying with temperature due to the strong Fe–Fe exchange coupling (see Sect. 2.2). The slight differences in the slope of the curves for several compositions can be explained by different growth conditions during deposition caused by the varying amount of Tb in the alloy films. A variation based on the contribution of the shape anisotropy due to the net saturation magnetization is far too small to cover the observed differences.

In order to provide a summary of the magnetic properties of 20-nm-thick amorphous FI $Fe_{100-x}Tb_x$ alloy films an overview of the effective magnetic anisotropy and the magnitude of the net saturation magnetization is shown in Fig. 6.9a, b, respectively, as a function of the Tb content x and the temperature. Furthermore, the

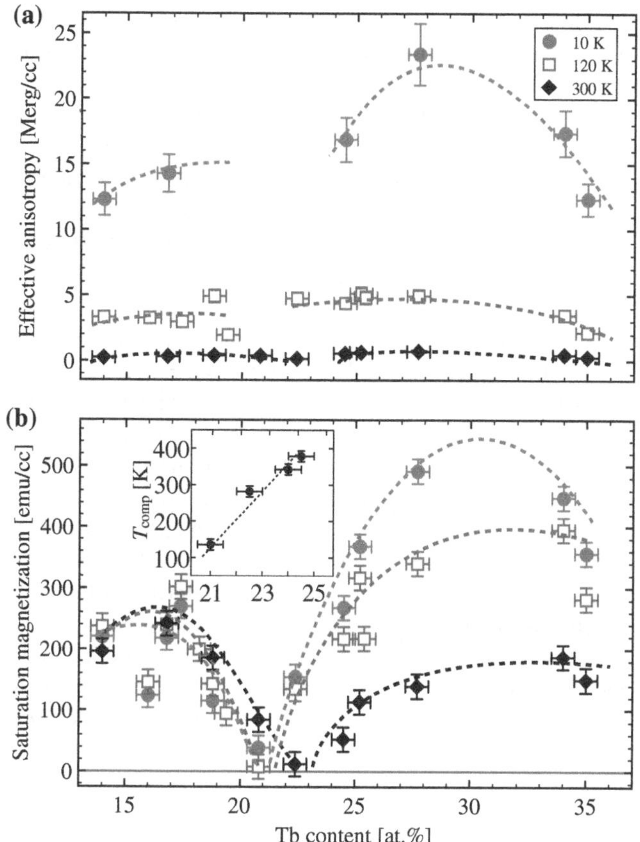

Fig. 6.9 **a** Effective magnetic anisotropy and **b** net saturation magnetization of 20-nm-thick amorphous $Fe_{100-x}Tb_x$ films as a function of the Tb concentration for temperatures of 10, 120, and 300 K. The *dashed curves* act as a guid to the eye. The discontinuity in the curves indicate the vicinity of the compensation point, where the exact value of the effective magnetic anisotropy is unknown. The inset in (**b**) shows the increase in compensation temperature towards higher amounts of Tb

composition dependence of the compensation temperature is given in the inset. The effective magnetic anisotropy, which was calculated by coercivity and net saturation magnetization as a lower estimate, reveals a strong temperature dependence, while the Tb content of the alloy has a weaker impact, which can be seen in Fig. 6.9a. Towards lower temperatures K_{eff} reaches values between 2 Merg/cc and 14 Merg/cc being relatively unaffected by the alloy composition except for high and low amounts of Tb, where K_{eff} tends to decrease. From RT measurements a slight decrease in anisotropy was also observed near compensation as discussed in Sect. 6.2.1. However, the exact anisotropy values in the vicinity of the compensation point at other temperatures are unknown due to the very strong coercivity enhancement and a change in the magnetization reversal process, which will be discussed later. The stoichiometry itself determines strongly the net saturation magnetization shown in Fig. 6.9b. Starting from the RT (300 K) compensation composition, the net saturation magnetization increases from zero towards lower and higher amounts of Tb. Thus, the magnetization compensation point depends strongly on the Tb content revealing a compensation temperature between 100 and 300 K for a Tb content between 20 and 24 at.%, respectively, as presented in the inset in Fig. 6.9b. Generally, one assumes that the observed temperature dependence of K_{eff} and the net saturation magnetization is attributed to the sperimagnetic nature of the amorphous Fe–Tb alloy system [17, 20, 21]. The Tb atoms possess a magnetic moment of about 9 μ_B, which is hardly changing with temperature [17]. However, the magnetic configuration in such a thin alloy film exhibits an orientational distribution of the magnetic moments, which determines the total magnetization of the Tb sublattice. This non-collinearity results from the equilibrium between the exchange interaction of the Fe–Fe, Fe–Tb, and Tb-Tb pairs and the single ion anisotropy of the Tb atoms caused by the interplay of their 4f orbitals with the local crystal field in the amorphous alloy. Since the orientational distribution of the Tb moments varies with temperature also K_{eff} and the net saturation magnetization changes.

The particular structure of the Tb fanning cone determines the average integral magnetic properties of the FI Fe–Tb films. In consequence the characteristic of the magnetization reversal process is influenced by the fanning cone and its temperature dependence. The following paragraphs deal with a detailed investigation of the magnetization reversal processes with regard to the temperature, dominant sublattice magnetization, and fanning cone structure.

6.3.1 Magnetization Reversal for Various Temperatures in Fe–Tb Films Dominated by the Fe Sublattice

The investigation of the RT magnetization reversal behavior of thin amorphous Fe–Tb films with respect to the alloy composition discussed in Sect. 6.2 revealed a strong relation to the dominant sublattice magnetization and particular structure of the orientational distribution of the Tb moments leading to distinct domain configurations and reversal mechanisms. Since the FI configuration of the Fe–Tb alloy is strongly influenced by the temperature, changes in the magnetization reversal can be expected in the range of $0 < T \leq T_C$ with $T_C \approx 400$ K.

Figure 6.10 shows the magnetization hysteresis loops and the first derivative of the left reversal branch of a 20-nm-thick $Fe_{83}Tb_{17}$ film at various temperatures as an example for the reversal of Fe sublattice dominated alloy films. The magnetization reversal in amorphous Fe–Tb films with a net saturation magnetization dominated by the Fe sublattice is strongly determined by the interaction of the external magnetic field and the collinear Fe moments. In this manner the reversal at 300 K occurs within a single sharp step indicating switching via domain nucleation and wall

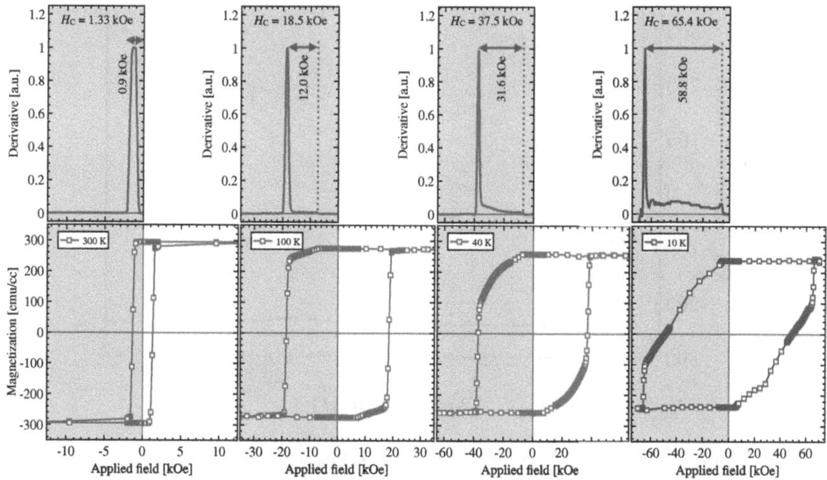

Fig. 6.10 Magnetization hysteresis loops of a 20-nm-thick $Fe_{83}Tb_{17}$ alloy film obtained from SQUID magnetometry measurements in out-of-plane geometry at various temperatures. And derivative of the reversal branch in the negative field range of the measured hysteresis loops indicated with the grey background. The center of the peak gives the coercive field for the magnetization reversal and the range with a derivative different from zero represents the width of the reversal branch

propagation as previously discussed in Sect. 6.2.2. Towards lower temperatures the coercivity increases due to the strong anisotropy enhancement. Furthermore, a significant change of the magnetization reversal sets in below 100 K. In this temperature region the nucleation field and the coercive field become separated, which manifests itself in a strong broadening of the reversal branch. With decreasing temperature from 100 K to 10 K the width of the reversal region increases from about 12 to 59 kOe, respectively (see the upper diagrams in Fig. 6.10). This is attributed to the coercivity enhancement, while the nucleation field itself remains more or less constant.

To elucidate the characteristic of the magnetization reversal a minor loop reversal study was performed on a 20-nm-thick $Fe_{83}Tb_{17}$ alloy film at 10 K. The magnetization reversal curves were obtained by saturating the sample in a positive field of 70 kOe, applying the corresponding reversal field, and measuring the magnetization during the external field is reduced to zero. The minor loops according to the different applied reversal fields are presented together with the full loop in Fig. 6.11. Reversal fields of up to 30 kOe yield a complete reversible field dependence of the Fe–Tb films magnetization as can be seen from the closed loops and the full remanent value at zero field. Higher fields lead to non-reversible magnetization field characteristics. However, as far as the reversal field remains below 60 kOe the magnetization reaches almost its full remanent value at zero field. A further increase of the reversal field in the range of the coercivity causes a significant modification of the minor loop. Although the magnetization increases again with reducing field, more than

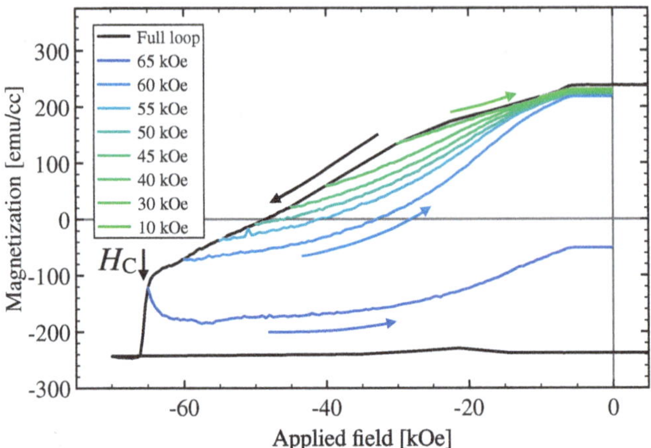

Fig. 6.11 Minor loop reversal study of a 20-nm-thick $Fe_{83}Tb_{17}$ alloy film measured in out-of-plane direction at 10 K by SQUID magnetometry. Please note, the reversal curves were obtained by saturating the sample in a positive field of 70 kOe, applying the corresponding reversal field, and measuring the magnetization during the external field is reduced to zero

half of the magnetic film has switched. Please note that the reduction of magnetization in the vicinity of the coercive field is related to some domain dynamics during the irreversible switching process. The observed minor loop characteristic allows the assumption that the magnetization reversal starts by nucleation of lateral domains. According to the literature [14] these domains possess a bubble like shape as discussed in Sect. 6.2.2. Therefore, the reversible part of the switching branch can be attributed to the expansion or contraction of the nucleated bubble domains with increasing or decreasing external field, respectively. The irreversible part and in particular the coercive field is correlated to that field range where the several bubbles merge. Since the energy consumption for the bubble domain expansion is related to the magnetic anisotropy energy of the material [22], the increasing PMA of Fe–Tb alloy films towards lower temperatures leads to higher magnetic fields required for the domain growth. As a consequence the reversal branch broadens, which can be seen in Fig. 6.10.

6.3.2 Magnetization Reversal for Various Temperatures in Fe–Tb Films Dominated by the Tb Sublattice

The fanning cone structure of the Tb sublattice moments has a strong impact on the interaction between the external field and the magnetic configuration of the Fe–Tb alloy films, in particular if the net saturation magnetization is dominated by the Tb sublattice. The temperature dependence of the out-of-plane magnetization hysteresis

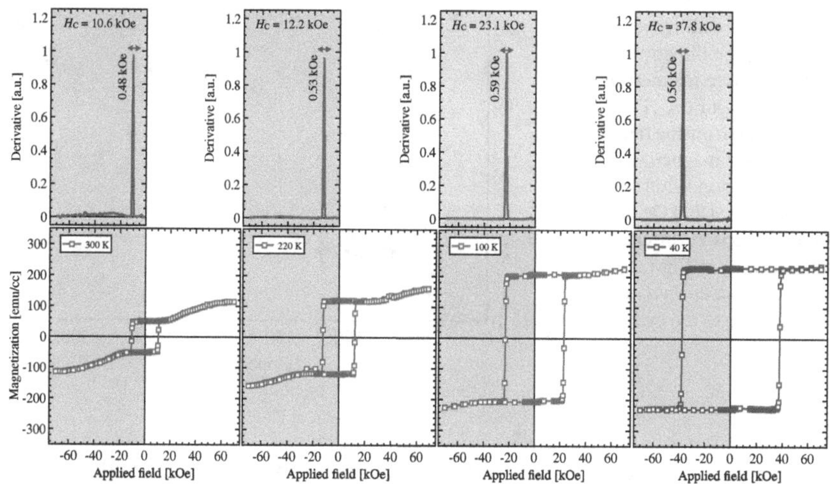

Fig. 6.12 Magnetization hysteresis loops of a 20-nm-thick $Fe_{75.5}Tb_{24.5}$ alloy film obtained from SQUID magnetometry measurements in out-of-plane geometry at various temperatures. The reversible features in the high field range can be attributed to the particular sperimagnetic configuration of the alloy. The derivative of the reversal branch in the negative field range of the measured hysteresis loops indicated with the grey background are presented in the upper diagrams. The center of the peak gives the coercive field for the magnetization reversal and the full width at half maximum represents an estimate of the width of the reversal branch

curves of a 20-nm-thick $Fe_{75.5}Tb_{24.5}$ alloy film as well as the first derivative are presented in Fig. 6.12.

As already discussed in Sect. 6.2.3 the hysteresis loops of Tb dominated Fe–Tb films exhibit a sharp reversal branch, which is nearly unaffected by the Tb content and the net saturation magnetization. Other temperatures yield similar characteristics. With decreasing temperature the net saturation magnetization as well as the coercivity increase. However, the width of the reversal branch remains more or less constant, as can be seen from Fig. 6.12. This allows the assumption that the magnetization reversal occurs via domain nucleation and wall propagation independently from the temperature. In addition to the main irreversible switching step the magnetization increases further towards higher fields starting at about 20 kOe. The enhancement is completely reversible and the relative change becomes reduced with decreasing temperature and increasing net saturation magnetization. This behavior originates most likely from the interaction between the external field and the orientational distribution of the Tb sublattice moments as discussed for the RT magnetization reversal in Sect. 6.2.3. The strong external magnetic field compresses the fanning cone structure of the Tb moments leading to higher magnetization values along the out-of-plane direction.

According to the macro-spin model, where the moment distribution becomes approximated by a single canted spin, the average zero field opening angle of the

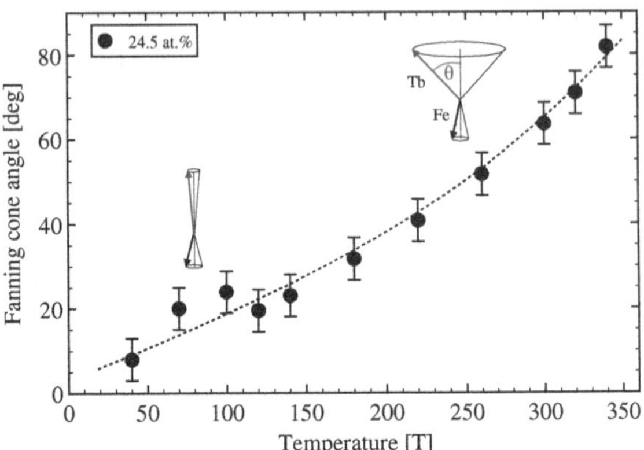

Fig. 6.13 Fanning cone opening angle with respect to the out-of-plane direction for the Tb sublattice magnetization present in a $Fe_{75.5}Tb_{24.5}$ film estimated from the ratio of the remanence magnetization and the saturation value of the reversible part at 70 kOe in the hysteresis loops at different temperatures (see Fig. 6.12). Please note, the *dashed curve* acts as a guide to the eye

fanning cone can be estimated with Eq. 6.1 by using the remanence and saturation values of the magnetization. Figure 6.13 presents the calculated opening angle of the fanning cone with respect to the out-of-plane Tb sublattice magnetization as a function of the temperature of a 20-nm-thick $Fe_{75.5}Tb_{24.5}$ film. A decrease in temperature from 340 to 40 K reduces the average opening angle from about 80° to about 10°, respectively. This narrowing of the fanning cone is caused by the enhancement of the magnetic exchange coupling towards lower temperatures and is accompanied by a strong increase in the effective magnetic anisotropy and net saturation magnetization in out-of-plane direction.

6.4 Reversal Mechanism in the Vicinity of the Compensation Point

The compensation point in FI film systems represents a particular point, where the magnetic configuration of the film is nearly uninfluenceable by the external magnetic field due to the compensated sublattice moments. Only within strong magnetic fields so called spin-flop or spin-flip transitions can occur changing the orientation of the magnetic moments as known commonly from AF materials [23–27]. A spin-flop describes the irreversible rotation of the antiparallel aligned sublattice moments towards the field direction leading to a transition from the AF into the F state. This can result in a canted spin structure. A continuous transition from the canted into the full parallel F configuration of the magnetic moments is often referred as spin-flip [28]. Depending on the moment configuration, the magnetic anisotropy, and the exchange coupling such transitions take place only in relative high magnetic fields, as will be discussed in the following.

Contrary to crystalline AF films amorphous FI films exhibit a orientational distribution of the magnetic moments in the sublattices, which is present even at the compensation point. As previously discussed, this magnetic moment distribution determines

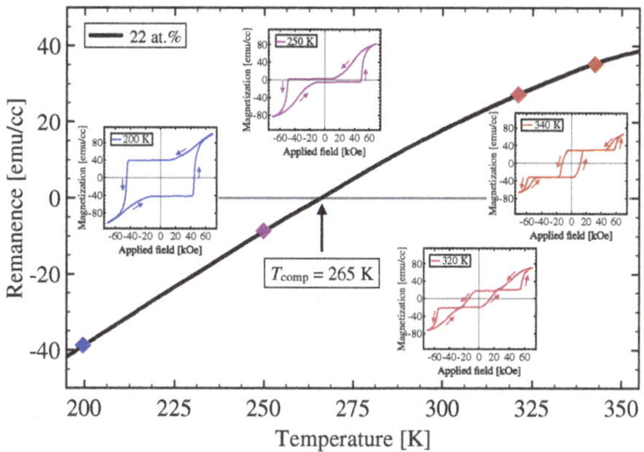

Fig. 6.14 Temperature dependence of the net magnetization in remanence state of a 20-nm-thick $Fe_{78}Tb_{22}$ film in the vicinity of the compensation point $T_{comp} = 265$ K. Several insets present the magnetization hysteresis loops at different temperatures indicating a strong change in the reversal mechanism with respect to the temperature and net magnetization

strongly the interaction between the external field and the configuration of the film. In consequence one can expect manyfold magnetic configurations in amorphous FI films within the vicinity of their compensation point depending on the external field strength. To elucidate this, amorphous Fe–Tb alloy films were investigated by SQUID magnetometry and element specific XMCD absorption measurements near the compensation point. Figure 6.14 shows the net magnetization in remanence state as a function of the temperature of a 20-nm-thick $Fe_{78}Tb_{22}$ film near its compensation point at $T_{comp} = 265$ K. The magnetization hysteresis loops presented in insets reveal a strong change in the reversal mechanism with respect to the temperature and net magnetization.

Starting from higher temperatures (e.g., at 340 K) the field dependence of the magnetization is characterized by a center and two satellite hysteresis loops. The center hysteresis loop corresponds to the magnetization reversal driven by the interaction of the external magnetic field and the dominant Fe sublattice moments. In the high field regime above 40 kOe the satellite hysteresis loops originate most likely from an additional spin reorientation transition as referred to a spin-flop or spin-flip. This might be related to a change in the fanning cone structure of the Tb sublattice moments forced by the oppositely oriented high magnetic field. In this particular case such a transition requires a magnetic field higher than 60 kOe, as the Zeeman energy needs to overcome the intrinsic exchange energies between Fe–Fe, Fe–Tb, and Tb-Tb sites in order to modify the equilibrium orientational moment distribution (see Sect. 2.2). With decreasing temperature towards the compensation point the net saturation magnetization becomes reduced and the center hysteresis loop vanishes (see the progress e.g., from 320 to 250 K). Very close to the compensation point a reversal takes place only due to the field induced spin-flop transition, as the field dependence of the magnetization yield only satellite hysteresis loops. Below compensation the net magnetization increases again due to the gain in magnetic moment of the now dominant Tb sublattice. Thus, the spin-flop transition becomes superimposed by the normal switching process of the magnetization indicated by the single reversal branch including both processes (e.g., at 200 K).

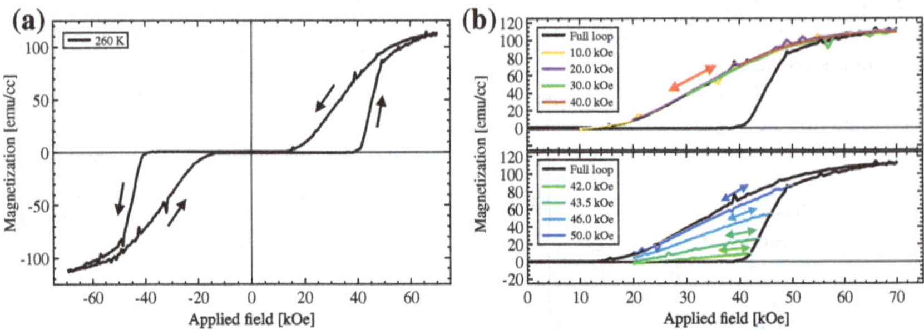

Fig. 6.15 **a** Magnetization hysteresis loop of a 20-nm-thick $Fe_{78}Tb_{22}$ film at 260 K close to the compensation temperature. **b** Minor loop reversal study obtained after saturation in an external magnetic field of 70 and −70 kOe shown in the upper and lower diagram, respectively. The minor loops after saturation at 70 kOe were measured with increasing magnetic field starting at different field values as designated in the legend. For the minor loop study after negative field saturation with −70 kOe a positive reversal field was applied and the magnetization was recorded with decreasing external field

The peculiar shape of the satellite hysteresis loops provide information about the characteristic of the spin-flop transition in the high field regime. Although the onset field for the sharp reversal branch and the enclosed area of the hysteresis changes with temperature, the shape stays more or less the same. To investigate the process properties of the spin-flop transition, a minor loop reversal study very close to the compensation point (e.g., at 260 K) of a 20-nm-thick $Fe_{78}Tb_{22}$ film was performed. The complete magnetization reversal loop as outlined in Fig. 6.15a reveals a nearly zero remanence state and the fin shaped satellite hysteresis loops in the high magnetic field range. The onset of the sharp reversal branch lies above 40 kOe, where the magnetization increases strongly up to 50 kOe. A further field enhancement towards 70 kOe leads to a somewhat smaller increase in magnetization. Starting from nearly saturation and reducing the external field the loop opens again at about 50 kOe and the magnetization decreases more or less slowly and reaches zero at about 20 kOe. This behavior indicates that two different processes are involved in the magnetization reversal. The minor loops shown in Fig. 6.15b confirm this assumption. The upper diagram presents the progress of the magnetization after the Fe–Tb film became saturated in a positive field of 70 kOe. After reducing the external field to a certain value and increasing it again up to 70 kOe the progress in magnetization is identical independent from the minimum field value. Thus, the change in magnetization after saturation is fully reversible and domain formation can be excluded as mechanism for the reversal. Contrary to this the behavior becomes different after saturating the sample in a negative field of −70 kOe, as can be seen from the lower diagram in Fig. 6.15b. Depending on the strength of the maximum positive applied field, the magnetization follows different reversal curves, when the field becomes cycled between 20 kOe and several maximum field values. Based on this results the sharp reversal branch of the satellite hysteresis loops is characterized by an irreversible switching process accompanied by the formation of magnetic domains.

According to the temperature dependence of the magnetization hysteresis loops and the minor loop reversal study in the vicinity of the compensation point the broad and the sharp switching branches of the satellite hysteresis loops of the amorphous Fe–Tb film are attributed to a reversible and an irreversible spin reorientation in the Fe and Tb sublattices, respectively. With regard to the sperimagnetic nature of the amorphous Fe–Tb alloy films the particular transition behavior is determined by the intrinsic configuration of the magnetic sublattice moments, as the fanning cone structure plays a crucial role for the interaction of the magnetic film with the external field, especially near the compensation point. In order to get detailed information about the sublattice configuration during the spin-flop transition, field dependent XMCD spectra deduced from the Fe $L_{2,3}$ and Tb $M_{4,5}$ absorption edges were obtained from a 20-nm-thick amorphous $Fe_{73.5}Tb_{26.5}$ film deposited on a 200-nm-thick Si_3N_4 membrane embedded in Pt seed and capping layers. Based on the XMCD signal the effective magnetic moment per atom projected on the out-of-plane direction can be calculated as a function of the external magnetic field for the Fe and Tb sublattices following the sum rules described in Sect. 5.4.2.

The amorphous $Fe_{73.5}Tb_{26.5}$ alloy film on the Si_3N_4 membrane reveals a compensation point slightly above 280 K and its magnetic properties in particular the magnetization hysteresis loops follow a similar temperature dependence as obtained for the $Fe_{78}Tb_{22}$ film on a standard Si wafer. In general one expects a strong variation in the magnetic behavior due to the different amount of Tb in the alloy (see Sect. 6.2.1). An obvious reason for the similarity between both films could be a sufficiently high amount of oxygen in the $Fe_{73.5}Tb_{26.5}$ film reducing the number of magnetically active Tb atoms. However, with regard to the zero field absorption and XMCD spectra in Fig. 6.16 no additional shoulder caused by a metallic oxide is present in the vicinity of the absorption edges. The spectra exhibit a clear metallic behavior for Fe and Tb with the characteristic fine structure known from literature [29–32]. Therefore, a significant influence on the magnetic properties of the FI alloy film due to the presence of oxygen can be neglected.

As an influence on the magnetic properties of the Fe–Tb alloys due to oxygen is ruled out, the similarity between both films with 22 and 26.5 at.% manifests itself most likely in different growth conditions on Si substrates and Si_3N_4 membranes, respectively. Thus, a variation in the fanning cone structure can occur, which determines the temperature behavior of the magnetic properties and particularly the compensation point. This allows in such an individual case similar magnetic properties despite differences in the alloy stoichiometry.

The projected out-of-plane Fe and Tb sublattice moment per atom of the $Fe_{73.5}Tb_{26.5}$ film obtained as a function of the applied magnetic field from XMCD absorption measurements at 280 K and the corresponding magnetization hysteresis loop received from SQUID magnetometry is shown in Fig. 6.17a, b, respectively. The inset in (b) presents the field dependence of the net moment calculated from the element specific hysteresis loops in (a) by taking into account the stoichiometry of the alloy film. The progress of the net moment is in good agreement to the SQUID hysteresis loop. Furthermore, schemes 1–3 provide a qualitative illustration of the average fanning cone structure within different external fields (marked in the

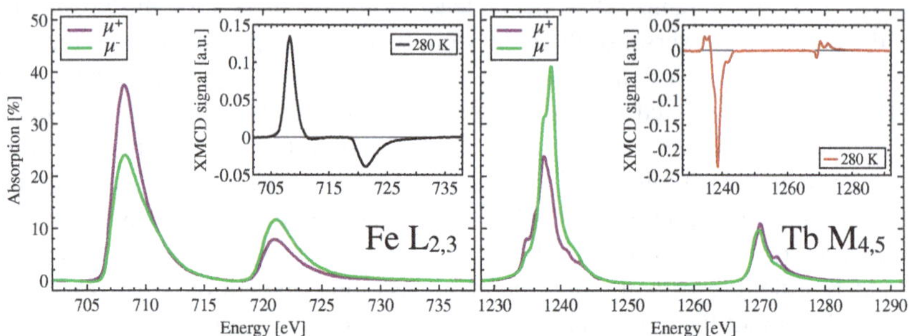

Fig. 6.16 X-ray absorption spectra at the Fe $L_{2,3}$ and Tb $M_{4,5}$ edges of a 20-nm-thick $Fe_{73.5}Tb_{26.5}$ alloy film on a 200-nm-thick Si_3N_4 membrane obtained in transmission geometry using circularly polarized X-rays with positive and negative helicity. The insets present the XMCD spectra calculated from the absorption spectra for different incident beam helicities. Please note, the spectra were measured in zero external magnetic field after saturating the sample in a positive field of 60 kOe

hysteresis loops) of the Fe and Tb sublattices assuming a homogeneous moment distribution. As mentioned above the magnetic moment per atom of the Fe and Tb sublattices was calculated using sum rules described in Sect. 5.4.2. In this manner the total moment results from the z-component of the orbital and spin moment. Concerning the spin moment calculation one needs to account for the magnetic dipole operator $\langle T_z \rangle$. Based on the literature [33, 34] the value of the dipole operator for the transition-metal atoms is nearly zero and can be neglected. For the rare-earth atoms atomic calculations published by Y. Teramura et al. [35] provide a sufficiently accurate estimation of the ratio between $\langle T_z \rangle$ and the spin moment $\langle S_z \rangle$ of -0.08. This leads to a correction factor of $1/0.76$ for the integral in the sum rule of the Tb spin moment (see Eq. 5.16). Taking into account the estimated magnetic dipole operator and a polarization degree of about 90 % for the incident X-rays during the absorption measurements, the experiment provides an accuracy of about 20 % for the calculation of the magnetic moment per atom of the Fe and Tb sublattices.

The element specific hysteresis loops in Fig. 6.17a reveal an antiparallel alignment between the Fe and Tb moments as expected for the FI Fe–Tb films. In zero field a Fe moment of about 1 μ_B/atom and a Tb moment of about -2.5 μ_B/atom were found (see point 1). These values are significantly smaller compared to the literature values [17] of about 2 μ_B/atom and 9 μ_B/atom for the Fe and Tb, respectively. This is a strong indication for the sperimagnetic nature of the Fe–Tb alloy films and the existence of an orientational moment distribution in the Tb sublattice and in particular also in the Fe sublattice (see scheme 1 in Fig. 6.17). After negative saturation, the magnetization at point 1 is negative, as slightly below the compensation point the Tb sublattice magnetization is already dominant. With increasing external magnetic field up to 40 kOe the Fe and Tb sublattice moments stay constant similar to the total net magnetization in Fig. 6.17b. Further field enhancement has a strong impact on the magnetic configuration. At about 48 kOe the Tb moments start to rotate into the applied field direction, while the Fe moments rotate *vice versa*. The change of the

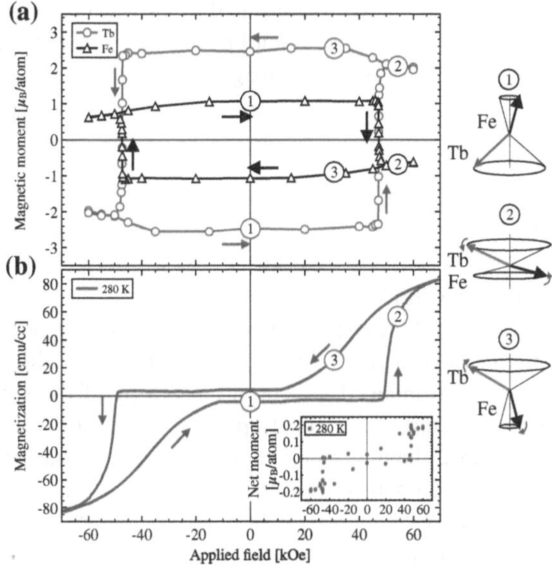

Fig. 6.17 **a** Element specific hysteresis loops of the projected out-of-plane Fe and Tb sublattice moment per atom of a 20-nm-thick $Fe_{73.5}Tb_{26.5}$ film at 280 K in the vicinity of the compensation temperature calculated from XMCD spectra at the Fe $L_{2,3}$ and Tb $M_{4,5}$ absorption edges. **b** Field dependence of the average net magnetization obtained from SQUID measurements. The inset shows the effective net magnetic moment per atom deduced from the element specific loops in (**a**) by taking into account the stoichiometry of the amorphous Fe–Tb alloy. The schemes 1–3 provide a qualitative illustration of the average fanning cone structure within different external fields of the Fe and Tb sublattices assuming a homogeneous moment distribution

Fe moments occurs at slightly higher magnetic fields compared to the reversal of the Tb moments. Within this irreversible switching process the orientational distributions of the Tb and Fe sublattice moments become modified in a spin-flop like manner. Towards even higher fields than 60 kOe (see point 2) the net magnetization still increases, which can be attributed to a reversible change of the fanning cone as illustrated in scheme 2. The strong external field forces a rotation of the Fe moments into the field direction indicated by the reduction of the average z-component. Simultaneously, the Tb moments rotate against the field driven by the antiparallel exchange coupling to the Fe moments. The particular structure of this high field state is determined by the competition between exchange energy and Zeeman energy. When the field becomes reduced again the magnetic moments of both sublattices relax into the equilibrium state, where the Tb fanning cone points in positive field direction as outlined in scheme 3.

According to the minor loop SQUID study on the $Fe_{78}Tb_{22}$ film at 260 K (see Fig. 6.15) the process of relaxation reveals a reversible characteristic originating from the sperimagnetic nature of the amorphous Fe–Tb alloy system. Based on this results the sharp irreversible reversal branch and the rather broad reversible branch in the hysteresis loop of Fig. 6.17b are attributed to a spin-flop reversal followed by a relaxation of the moments. Such a peculiar behavior was predicted by G. V. Sayko et al. [36] for sperimagnetic films in high magnetic fields, especially in the vicinity of the compensation point.

6.5 Summary

The amorphous structure of thin $Fe_{100-x}Tb_x$ alloy films and the antiparallel exchange coupling between Fe and Tb moments give rise for a sperimagnetic configuration with strong sensitivity to growth conditions, temperature, and alloy stoichiometry. In this manner the growth induced structural anisotropy of the short range order in conjunction with the single ion anisotropy determines the sperimagnetic spin configuration (see Chap. 2) having a noticeable influence on the integral magnetic properties like net saturation magnetization, coercivity, and magnetic anisotropy.

Beside the commonly known [4] stoichiometry dependent changes of the net magnetization and coercivity at RT strong variations in the reversal process for Fe and Tb dominated alloy films were found mainly in the vicinity of the compensation point. The reducing net magnetization with increasing Tb content towards the compensation point at 23 at.% in Fe dominated Fe–Tb films leads to a broadening of the reversal branch, which occurs most likely due to a change in the reversal process from stripe to bubble domain reversal, as observed already in thicker films [14, 15]. Terbium dominated films behave differently near the compensation point. While the irreversible reversal branch is nearly unaffected by the stoichiometry except of a coercivity enhancement towards the compensation point, an additional reversible increase in the net magnetization appears within high magnetic fields above 20 kOe. This is attributed to the interaction between the external field and the orientational moment distribution (fanning cone) of the Tb sublattice in the sperimagnetic Fe–Tb film system. Large applied magnetic fields lead to a reversible decrease of the average fanning cone opening angle, which is a peculiarity of the sperimagnetic configuration.

As the magnetic exchange coupling between the Fe–Fe, Fe–Tb, and Tb-Tb sites determines the fanning cone structure of the sublattice moments, the gain in exchange energy towards lower temperatures leads to a strong enhancement of the Tb sublattice magnetization because of a narrowing of the average fanning cone opening angle, which yield a higher projected moment along the out-of-plane direction. This gives rise for the existence of a compensation point in a composition range of 20 at.% $< x \leq$ 24 at.%, where the Tb sublattice magnetization overcomes that of the Fe sublattice. Please note that the particular stoichiometry range and compensation temperature depend strongly on the film growth conditions. A further consequence of the exchange energy enhancement is the increase of coercivity driven by the magnetic anisotropy, which is also directly related to the narrowing of the fanning cone structure. This occurs more or less independently from the alloy composition.

Similar to the results obtained at RT the orientational moment distribution of the magnetic sublattices influences the reversal process of the net magnetization.

A decreasing temperature leads to large broadening of the reversal branch in Fe dominated Fe–Tb films due to a strong coercivity enhancement, while the domain nucleation field remains more or less constant. This can be attributed to the nucleation of lateral domains with a bubble like shape nucleating in the lower field range and expanding reversible up to the coercive field, where the domains merge, and the full reversal becomes completed. Contrary to this, Tb dominated Fe–Tb films reveal no significant change in the magnetization reversal branch towards lower temperatures. However, the additional reversible increase in net magnetization within higher magnetic fields vanishes with decreasing temperature. With respect to the fanning cone structure of the Tb sublattice moments this arises from the narrowing of the average opening angle and increasing PMA. In particular a reduction of the opening angle from about 80° to about 10° from 340 to 40 K was found.

The intrinsic magnetic configuration of the sperimagnetic Fe–Tb alloy system possesses a strong temperature and stoichiometry dependence as well as a peculiar field dependency specifically in the vicinity of the compensation point, where the net magnetization vanishes. A modification of the orientation of magnetic moments is only driven by a spin-flop transition in the high magnetic field regime. In this manner satellite hysteresis loops consisting of a sharp irreversible reversal branch with increasing field and a rather broad reversible branch with decreasing field were observed near the compensation point. Based on SQUID minor loop studies and element specific hysteresis loops obtained from XMCD absorption measurement the mechanisms attributed to the particular characteristic of the satellite hysteresis loops could be identified as a spin-flop reversal followed by a relaxation of the magnetic moments. This behavior was predicted by G. V. Sayko et al. [36] as a characteristic for the sperimagnetic nature of amorphous FI heavy rare-earth-transition-metal films.

References

1. M.A. Phillips, V. Ramaswamy, B.M. Clemens, W.D. Nix, J. Mater. Res. **15**, 2540 (2000)
2. S. Wei, B. Li, T. Fujimoto, I. Kojima, Phys. Rev. B **58**, 3605 (1998)
3. H. Schletter, Präparation und Charakterisierung nanostrukturierter Magnetwerkstoffe unter besonderer Berücksichtigung des Exchange Bias Effekts, Ph.D. Thesis, TU Chemnitz (2013)
4. Y. Mimura, N. Imamura, Appl. Phys. Lett. **28**, 746 (1976)
5. Y. Mimura, N. Imamura, T. Kobayashi, A. Okada, Y. Kushiro, J. Appl. Phys. **49**, 1208 (1978)
6. R. Van Dover, M. Hong, E. Gyorgy, J. Dillon, S. Albiston, J. Appl. Phys. **57**, 3897 (1985)
7. Y. Mimura, N. Imamura, T. Kobayashi, IEEE Trans. Magn. **12**, 779 (1976)
8. C. Robinson, M. Samant, Appl. Phys. A Mater. Sci. Process (1989)
9. M. Tewes, J. Zweck, H. Hoffmann, J. Magn. Magn. Mater. **95**, 43 (1991)
10. V.G. Harris, K.D. Aylesworth, B.N. Das, W.T. Elam, N.C. Koon, Phys. Rev. Lett. **69**, 1939 (1992)
11. T. Hufnagel, S. Brennan, P. Zschack, B. Clemens, Phys. Rev. B **53**, 12024 (1996)
12. A. Kirilyuk, J. Ferré, V. Grolier, J.P. Jamet, D. Renard, J. Magn. Magn. Mater. **171**, 45 (1997)
13. J. Pommier, P. Meyer, G. Pénissard, J. Ferré, P. Bruno, D. Renard, Phys. Rev. Lett. **65**, 2054 (1990)
14. B. Lanchava, H. Hoffmann, J. Phys. D: Appl. Phys. **31**, 1991 (1998)

15. J. Schmidt, G. Skidmore, S. Foss, E. Dan Dahlberg, C. Merton, J. Magn. Magn. Mater. **190**, 81 (1998)
16. J. Coey, J. Appl. Phys. **49**, 1646 (1978)
17. K. Handrich, S. Kobe, *Amorphe Ferro- und Ferrimagnetika* (Akademie, Berlin, 1980). ISBN 978-3-87664-044-0
18. I. Campbell, J. Phys. F: Met. Phys. **2**, L47 (1972)
19. M. Mansuripur, M. Ruane, IEEE Trans. Magn. **22**, 33 (1986)
20. J. Coey, J. Chappert, J. Rebouillat, T. Wang, Phys. Rev. Lett. **36**, 1061 (1976)
21. J. Rebouillat, A. Lienard, J. Coey, R. Arrese-Boggiano, J. Chappert, Physica B **86**, 773 (1977)
22. J. Nielsen, Annu. Rev. Mater. Sci. **9**, 87 (1979)
23. L. Néel, Annales de Physique **5**, 232 (1936)
24. N.J. Poulis, G. Hardeman, Physica **18**, 201 (1952)
25. C. Gorter, Rev. Mod. Phys. **25**, 332 (1953)
26. E.A. Turov, *Physical Properties of Magnetically Ordered Crystals* (Academic Press, New York, 1965). ISBN 978-0-127-04950-2
27. A. Bogdanov, U. Rößler, M. Wolf, K.H. Müller, Phys. Rev. B **66**, 214410 (2002)
28. U.K. Roessler, A.N. Bogdanov, arXiv preprint cond-mat/0605493 (2006)
29. J.B. Goedkoop, J.C. Fuggle, B.T. Thole, G. Van der Laan, G.A. Sawatzky, J. Appl. Phys. **64**, 5595 (1988)
30. M. Sacchi, R.J.H. Kappert, J.C. Fuggle, E.E. Marinero, Appl. Phys. Lett. **59**, 872 (1991)
31. W. O'Brien, B. Tonner, Phys. Rev. B **50**, 12672 (1994)
32. G. Van der Laan, E. Arenholz, Z. Hu, A. Bauer, E. Weschke, C. Schüssler-Langeheine, E. Navas, A. Mühlig, G. Kaindl, J. Goedkoop et al., Phys. Rev. B **59**, 8835 (1999)
33. M. Tanaka, T. Asahi, A. Agui, M. Mizumaki, J. Sayama, T. Osaka, J. Phys. D: Appl. Phys. **41**, 055003 (2008)
34. Y. Guan, Z. Dios, D.A. Arena, L. Cheng, W.E. Bailey, J. Appl. Phys. **97**, 10A719 (2005)
35. Y. Teramura, A. Tanaka, B.T. Thole, T. Jo, J. Phys. Soc. Jpn. **65**, 3056 (1996)
36. G.V. Sayko, S.N. Utochkin, A.K. Zvezdin, J. Magn. Magn. Mater. **113**, 194 (1992)

Chapter 7
Percolated $Fe_{100-x}Tb_x$ Nanodot Arrays: Exchange Interaction and Magnetization Reversal

As a future concept on the route towards ultra high density magnetic data storage, percolated perpendicular media (PPM) was introduced in 2006 [1–3]. Since thermal stability becomes a key issue when the size of the magnetic unit, representing the data bit, is reduced below a certain value [4], percolated media follows a completely different approach. The grains of the magnetic material are fully exchange coupled yielding a high thermal stability of the magnetic configuration. The magnetic domains used to store the bits are locally stabilized by pinning effects, if the magnetic thin film is riddled with periodic highly ordered pinning sites like nonmagnetic arrays of dots or antidots. Recently it was shown that this concept works well for F materials with PMA, when the size of the pinning sites remains in the range of the domain wall width [5–7].

In percolated systems the intrinsic exchange interaction, the magnetic anisotropy, and the saturation magnetization play an important role for the pinning effect of the domain wall (DW) on the nonmagnetic defect, which influences the magnetization reversal process. This leads to the fundamental question, how does the magnetic configuration behave within an external field in a percolated system exhibiting small intrinsic exchange coupling and saturation magnetization. One suitable material class to answer this question is the rare-earth-transition-metal alloy system. In particular, thin FI Fe–Tb alloy films in a composition range from 14 to 35 at.% Tb represent a good model system due to the amorphous structure preventing grain boundary effects, high PMA ($7 - 10$ Merg/cc), low net saturation magnetization (<300 emu/cc), and weak exchange coupling ($\approx 10^{-7}$ erg/cm) as characteristic for this alloy system [8].

In the following an analysis of the morphology, magnetization reversal processes, coupling phenomena, and local magnetic configuration appearing in percolated $Fe_{100-x}Tb_x$ nanodot arrays with a stoichiometry varying from 19 to 23 at.% Tb will be presented. The fabrication of the percolated nanodots was realized by co-deposition of Fe and Tb at RT onto pre-patterned Si(100) substrates using magnetron sputtering. The 20-nm-thick amorphous $Fe_{100-x}Tb_x$ films are embedded inPt

C. Schubert, *Magnetic Order and Coupling Phenomena*, Springer Theses,
DOI: 10.1007/978-3-319-07106-0_7,

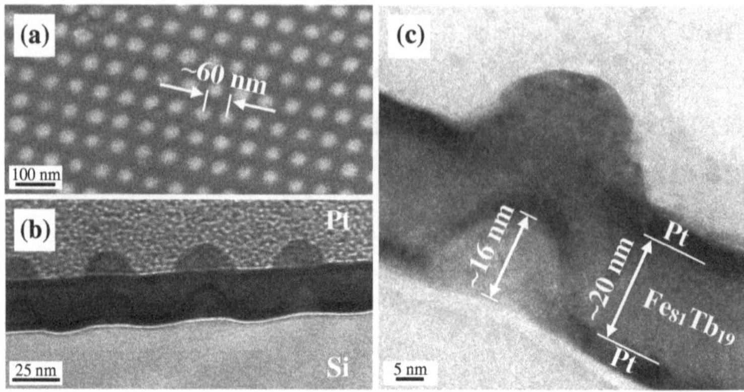

Fig. 7.1 **a** Topography image of Si nanodots of a pre-patterned substrate with a size of 30 nm and a pitch of 60 nm obtained from SEM investigations. The arrangement of the nanodots deviates from the hexagonal lattice. **b**–**c** TEM images with different magnification revealing a cross-sectional view of the $Fe_{81}Tb_{19}$ nanodots and trench material on top of dome like shaped substrate nanodots. The higher resolution image in (**c**) shows the layer stack following closely the topography of the substrate. Please note, the additional Pt layer on top of the presented layer stack originates from the TEM sample preparation by FIB

seed and capping layers with thicknesses of 5 and 3 nm, respectively, which prevent the Fe–Tb from oxidation. The pre-patterned substrates were fabricated by NIL as described in Sect. 5.2.

7.1 Morphology and Structural Properties of Thin Fe–Tb Films on Pre-patterned Substrates

The Si nanodots of the pre-patterned substrates possess a diameter of 30 nm, a height of about 16 nm, and a period of 60 nm. The SEM image in Fig. 7.1a shows the top view of the nanodot array. The nanodot arrangement differs from a hexagonal lattice. Overall the deviation in diameter, height, and period of the nanodots from the average values is rather small providing good substrate conditions for the deposition of percolated amorphous Fe–Tb films with highly ordered pinning sites.

The TEM images in Fig. 7.1b, c of a 20-nm-thick percolated $Fe_{81}Tb_{19}$ film reveal that the alloy films embedded in the Pt layers follow closely the morphology of the more of less dome shaped substrate pillars. The films can be seen as continuous, especially since no grain boundaries exist in the amorphous alloy. In this manner, the magnetic material on top of the pillars and in the trenches, in the following named as nanodots and trench material, is magnetically exchange coupled. This particular film morphology has strong impact on the magnetic properties and reversal behavior of the percolated Fe–Tb nanodots and trench material, which will be discussed in the following.

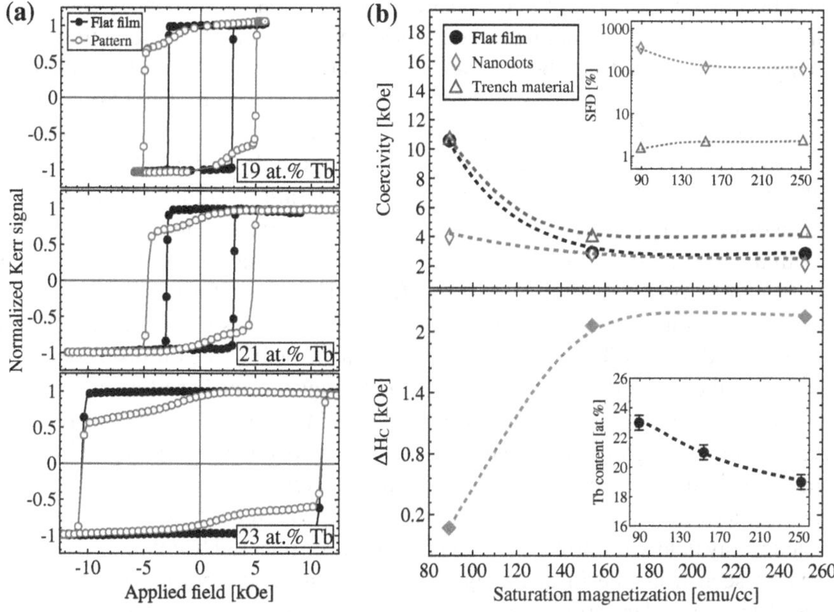

Fig. 7.2 **a** Normalized magnetization hysteresis loops of 20-nm-thick amorphous $Fe_{100-x}Tb_x$ alloy films on flat and pre-patterned substrates with a Tb content varying from 19 to 23 at.% measured using Kerr magnetometry at RT. **b** Coercivity of the Fe–Tb flat film, nanodots, and trench material, respectively (*upper diagram*) and difference in coercivity between flat film and trench material (*lower diagram*) as a function of the net saturation magnetization of the alloy system. The insets reveal the normalized SFD of the nanodots and the trench material according to the hysteresis loops in (**a**) as well as the relation between net saturation magnetization and Tb content of the alloy system. Please note, the *dashed curves* act as a guide to the eye

7.2 Magnetization Reversal and Pinning Effects

The integral magnetic properties of the $Fe_{100-x}Tb_x$ nanodots and trench material in comparison to the film on the flat region of the substrate, as presented in Fig. 7.2a, were investigated at RT using polar MOKE with a laser beam focused to a spot diameter of less than 200 μm, which allows to measure only the patterned area. The hysteresis loops of the percolated and flat films are shown for a Tb content of 19, 21, and 23 at.%, respectively.

Independent from stoichiometry and measurement area, whether on the flat or pre-patterned substrate, the Fe–Tb films exhibit PMA, which can be seen from the ratio of about one between the remanence and saturation magnetization. However, the Fe–Tb nanodots and trench material reverse separately contrary to the flat film, where the magnetization reversal takes place within a single step. The shoulders in the hysteresis loops obtained from the patterned area of the samples correspond to the nanodots reversal and reveal a broad switching field distribution indicating individual reversal events. In contrast to this the switching of the trench material is rather sharp and occurs at higher external fields. Pinning effects caused by magnetostatic interaction at the nonmagnetic pillars of the substrate can explain the strongly enhanced coercivity of the trench material as compared to the flat film observed in percolated Fe–Tb films with 19 and 21 at.% Tb. This is possible since the pillar size with a diameter of 30 nm and a height of 16 nm lies in the same order of magnitude as the DW width in the Fe–Tb alloy system, being typically in the range of 10 nm [5, 9]. In consequence the reversal of the trench material is dominated by the depinning field.

With a Tb content of 23 at.% the magnetization reversal of the trench material and the flat film occurs at similar magnetic fields. The reason for this is attributed to the net magnetization of the Fe–Tb alloy decreasing towards the RT compensation point at around 23 at.% Tb, which leads to an enhancement of the intrinsic coercivity as shown in the upper diagram of Fig. 7.2b and discussed for flat films in Sect. 6.2. Therefore, the coercivity difference between the flat film and the trench material decreases (lower diagram) and the intrinsic coercivity overcomes the depinning field in the percolated film determining the magnetization reversal behavior. The decrease in net magnetization has another impact on the reversal process. Due to the reduced magnetic stray field, the normalized SFD for the trench material becomes also smaller as can be seen from the inset of Fig. 7.2b. Please note, the normalized SFD was deduced from the derivative of the magnetization curve as the ratio between the FWHM of the reversal peak and the peak center obtained from a gaussian fit.

Contrary to the trench material the normalized SFD of the nanodots becomes broader with increasing Tb content, which is against the expectation that a smaller stray field will reduce the SFD. An explanation could be that in the vicinity of the compensation point small stoichiometric variations in the material of several nanodots have a strong impact on the individual reversal processes leading to a broader field range, where the switching of the individual nanodots occurs, as compared to Fe–Tb nanodots with smaller amount of Tb and higher net saturation magnetization.

7.3 Lokal Magnetic Properties and Domain Configuration

Evidence for pinning effects can be found in the domain pattern in the demagnetized state of the percolated Fe–Tb alloy films obtained from MFM imaging on the prepatterned sample area. The MFM image in Fig. 7.3 presents the domain pattern of a percolated $Fe_{79}Tb_{21}$ film after sample demagnetization. It reveals domains in the micrometer regime with domain walls following the arrangement of the nanodots. These are attributed to the trench material. At the flat film (not shown) no comparable domain structure exists and the domain size lies in the range of millimeters. Hence, the smaller domain structure of the trench material appears to be stabilized by pinning effects due to the nanodots. Surprisingly, the magnetization of the Fe–Tb nanodots itself resides in single domain states (see the inset of Fig. 7.3). In general this is not observed for exchange coupled magnetic systems. However, the exchange coupling in Fe–Tb alloy films is rather weak [8] and this might give rise to the formation of

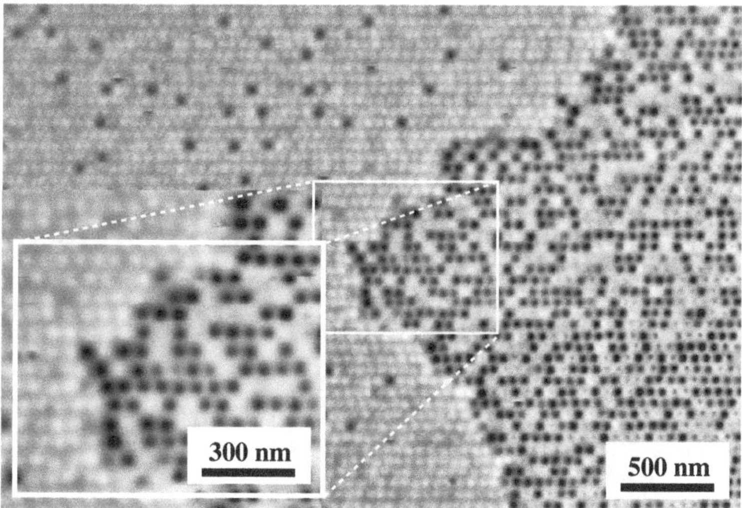

Fig. 7.3 Magnetic force microscopy image of a 20-nm-thick amorphous $Fe_{79}Tb_{21}$ film on top of a pre-patterned substrate in demagnetized state at RT. The inset provides an image with higher magnification showing the percolated Fe–Tb nanodots in magnetic single domain state

domains localized at the nanodots. Furthermore, the exchange interaction between the nanodots and the trench material manifests itself not solely in pinning, the MFM image shows also that more nanodots pointing parallel to the trench material as vice versa. Thus, the magnetic configuration of the trench material influences that of the nanodots, which has an impact on the overall magnetization reversal processes in these percolated Fe–Tb films.

In order to obtain detailed information about the magnetization reversal of the trench material a RT in-field MFM study was performed on a $Fe_{81}Tb_{19}$ nanodot array within an increasing external magnetic field starting from the demagnetized state. The domain configuration of the pattern are shown for different fields in Fig. 7.4a–i. In (a) the demagnetized state reveal the typical domain structure of the trench material where the domain walls follow more or less arbitrary the arrangement of the single domain nanodots. Within small fields in (b) and (c) this domain structure remains unchanged, while an increasing amount of nanodots become switched. First magnetization reversal events of the trench material occur not below 3 kOe as can be seen from (d). Compared to the zero field configuration in (a) the domain wall moved from the initial pinning site to another one, which is indicated with the red circle. From this point the reversal takes place in several switching events where the DW moves between different pinning sites (blue circle in (d) and (e)). Nevertheless, the field range of the reversal is rather narrow as the domain pattern vanishes and the trench material becomes saturated at 4 kOe (see Fig. 7.4d–i). Striking during this reversal process is the small domain size before saturation, which becomes stabilized due to the pinning to the nanodots.

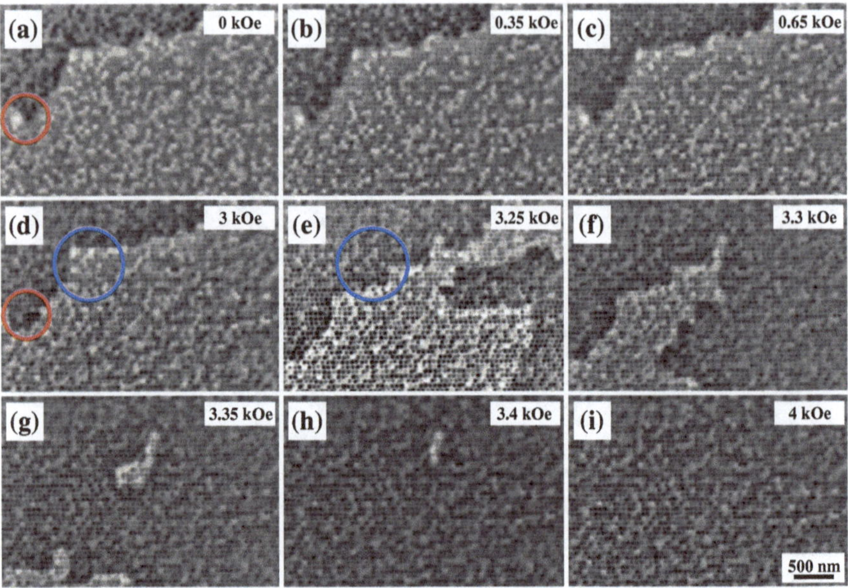

Fig. 7.4 **a–i** In-field MFM images within different applied magnetic fields starting from demagnetized state of a $Fe_{81}Tb_{19}$ nanodot array at RT. The magnetic signal originating from the single domain nanodots is superimposed by the domain structure of the trench material. The *red* and *blue circles* indicate the reversal of a domain by domain wall movement from one pinning site to another

According to the magnetization reversal behavior deduced from the MFM images strong domain wall pinning was observed in the percolated Fe–Tb film with a depinning field of about 3 kOe. Please note, this value differs slightly from the value of about 4 kOe obtained from the MOKE measurements presented in the upper diagram of Fig. 7.2a due to a deviation in the field calibration between the MOKE setup and in-field MFM.

The MFM images in Fig. 7.4 reveal a switching of the single domain Fe–Tb nanodots at smaller fields compared to the trench material as it is also observed in the MOKE study in Fig. 7.2. Additional to the previous investigation the nanodot switching within the domain pattern of the trench material was analyzed separately with increasing magnetic field after demagnetization. In particular only nanodots in a homogenous magnetized domain of the trench material were analyzed with the MFM. The in-field MFM images and magnetization curve are shown in Fig. 7.5a, b, respectively. The magnetization curve in (b) was calculated from the MFM images taking into account the number of dark and bright dots. In this manner black dots correspond to the positive field direction. Furthermore, the magnetic moment of the trench material domain points in negative field direction. In the demagnetized state more nanodots exhibit a bright domain, which originates from the exchange coupling between the nanodots and the trench material (compare Fig. 7.5a, b). With increasing field single domain switching events occur (see the red circles as guid to the eye) and

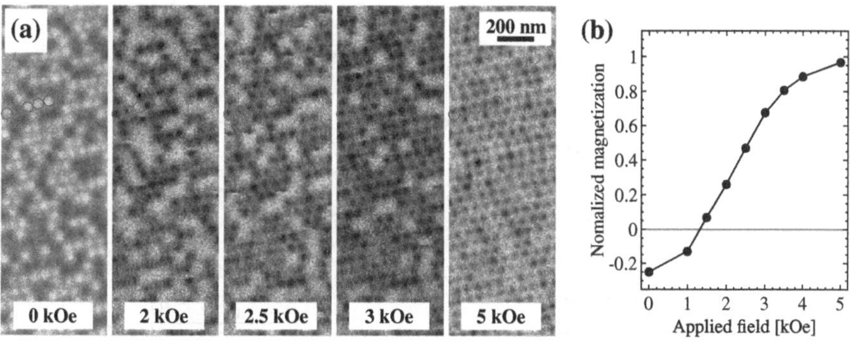

Fig. 7.5 **a** In-field MFM images and **b** initial magnetization curve of a $Fe_{81}Tb_{19}$ nanodot array measured with increasing external magnetic field at RT. The sample was initially demagnetized in an alternating magnetic field with decreasing amplitude. The magnetization curve in (**b**) was calculated from the MFM images taking into account the number of *dark* and *bright dots*. In this manner *black dots* correspond to the positive field direction. Please note, the *red circles* are intended as guide to the eye for the single nanodot switching events

the nanodots become saturated in positive field direction within a broader field range. The large SFD originates most likely from slight differences in the alloy properties of the Fe–Tb nanodots and a possible variation of the exchange coupling to the trench material. However, as soon as the trench material becomes reversed in a field of 5 kOe all nanodots point in positive field direction forced by the exchange coupling.

The observed local magnetization reversal behavior in the percolated Fe–Tb alloy system is strongly determined by the magnetic exchange coupling between the single domain nanodots and trench material. Nevertheless, the exchange coupling, which provides DW pinning for the trench material, does not hinder the occurrence of single domain magnetization states in the nanodots. To exclude magnetostatic interaction as origin for the observed peculiar magnetic configuration additional in-field and remanent Kerr hysteresis loops of a $Fe_{81}Tb_{19}$ nanodot array with trench material were obtained (see Fig. 7.6) at RT. Compared to the full hysteresis loop the direct current demagnetization (DCD) investigation, where the remanent Kerr rotation $M_R(H_r)$ was recorded after saturating the sample in a field of 20 kOe and applying a negative reversal field H_r, reveals no switching events as long as the reversal field reaches the coercivity of the trench material. From the full hysteresis loop one obtains that in remanence almost the same amount of nanodots exhibit a magnetization pointing in positive as well as negative field direction. According to magnetostatic interaction this is not favored since the magnetization of the trench material is completely oriented in positive field direction. In consequence and with respect to the DCD curve, magnetic exchange coupling between nanodots and trench material exists leading to the fact that the nanodots reverse back into their remanent configuration after applying a negative reversal field as long as the trench material is magnetized in positive field direction. This behavior is also found in the initial remanent magnetization (IRM) curve, which represents the remanent Kerr rotation with subsequent

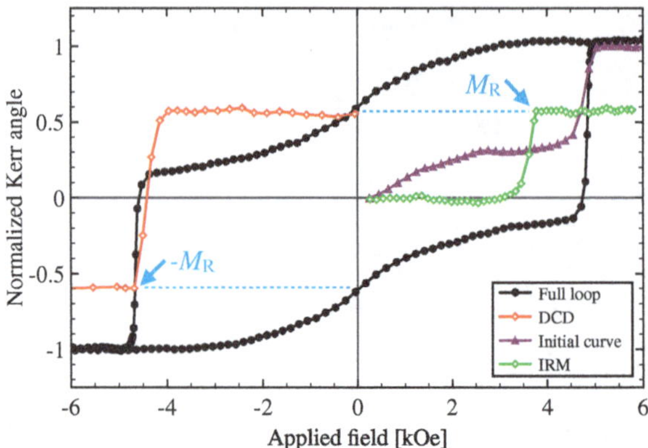

Fig. 7.6 Normalized Kerr hysteresis loops of a $Fe_{81}Tb_{19}$ nanodot array with trench material obtained from in-field and remanence measurements at RT. The initial curve was measured after demagnetizing the sample in an alternating external magnetic field with decreasing amplitude. For the DCD investigation the remanent Kerr rotation $M_R(H_r)$ was recorded after saturating the sample in a field of 20 kOe and applying a negative reversal field H_r. The IRM curve represents the remanent Kerr rotation with subsequent increasing external magnetic field after sample demagnetization. Please note, the *blue dashed lines* indicate the positive and negative remanence value of the full Kerr hysteresis loop

increasing external magnetic field after sample demagnetization. Starting from the demagnetized state the IRM curve remains at zero as long as the increasing external field reaches a value of about 3 kOe, while the initial curve reveals a noticeably enhancement of the normalized Kerr angle, which is attributed to the switching of the nanodots. Above 3 kOe the magnetization reversal of the trench material begins resulting in an increasing Kerr signal also for the IRM measurement. Please note, the deviation between the reversal fields of the initial curve and the IRM curve is caused by the many field cycles applied to the sample during the IRM measurement leading to field induced domain wall motion in the trench material, which results in a lower reversal field compared to the initial curve. Finally the observed results in particular from the DCD and IRM investigation lead to the assumption that the magnetic exchange dominates against magnetostatic interaction in the present percolated Fe–Tb alloy system.

7.4 Angular Dependency of the Switching Field for Nanodots and Trench Material

The investigation of the integral and local magnetic properties presented in this chapter reveals different reversal mechanism and magnetic domain structure for the $Fe_{100-x}Tb_x$ nanodots and the trench material, although the magnetic film on the pre-patterned substrate is fully exchange coupled. To elucidate the particular

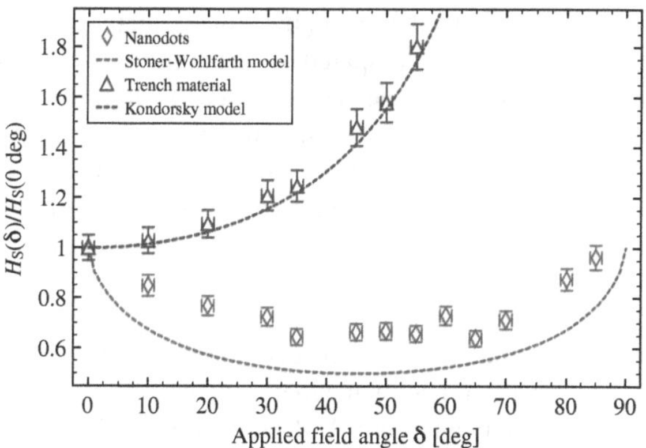

Fig. 7.7 Angular dependence of the switching field $H_S(\delta)$ for $Fe_{81}Tb_{19}$ nanodots and trench material obtained from angle dependent remanence Kerr hysteresis loops at RT. Please note that the switching field for the various angles is normalized to the switching field at zero degree in order to provide a better data view with respect to the Stoner-Wohlfarth and Kondorsky model

reversal mechanism of the nanodots and trench material, the angular dependence of the switching field was measured separately for both components using angle resolved MOKE remanence curves as described in Sect. 5.4.1.

The switching field $H_S(\delta)$ of the nanodots and trench material is shown in Fig. 7.7 as a function of applied field angle normalized to the switching field at zero degree. Please note that for the angle dependent study the direction of the external magnetic field was varied with respect to the sample normal, while the optical path of the laser beam was held in polar geometry. From the angular dependence of the switching field two different reversal processes can be deduced. While the nanodots come up with a more or less Stoner–Wohlfarth [10] like angular switching field dependence, the trench material follows a clear Kondorsky [11] behavior. Based on this and their single domain characteristic the reversal of the nanodots is most likely related to a nucleation-dominated process, which also explains the broad switching field distribution shown previously. In contrast to this the magnetization of the trench material reverses via DW motion. Strictly speaking the magnetization reversal of the percolated nanodot array occurs via individual nucleation of single domains located at the nanodots followed by a depinning of the already existing DWs from the nanodots and propagation through the trench material. These results are valid independent from the investigated stoichiometry of the percolated $Fe_{100-x}Tb_x$ films.

7.5 Summary

Percolated $Fe_{100-x}Tb_x$ alloy films with varying composition from 19 to 23 at.% Tb were fabricated using pre-patterned substrates. The highly ordered pre-patterned structure is completely transferred into the Fe–Tb layer forming nanodots, which are connected to the material in the trenches of the substrate. This provides full exchange coupling between the Fe–Tb nanodots and trench material. From Kerr hysteresis loops a separate switching of the nanodots and the trench material was observed. Independent from the stoichiometry an enhancement of the coercivity of the trench material compared to the flat film exists, which is attributed to DW pinning at the nanodots. For the percolated Fe–Tb film with a Tb content of 23 at.% near the RT compensation composition the intrinsic coercivity, enhanced by the decreasing net magnetization, overcomes the depinning field and determines the magnetization reversal. In-field MFM investigations reveal single domain nanodots and a domain structure of the trench material, which is mainly determined by DW pinning on the nanodots. In this context a depinning field of about 3 kOe was found. Furthermore, DCD and IRM measurements provide evidence that the magnetic domain configuration and magnetization reversal behavior of the percolated Fe–Tb alloy system is strongly determined by magnetic exchange coupling and that magnetostatic interaction plays only a minor role. Based on the angular dependence of the switching field a more or less nucleation-dominated reversal process occurs for the single domain nanodots, while the trench material reverses via DW motion after depinning from the nanodots. Finally the net magnetization in this percolated FI system plays a minor role for the pinning effects. However, in the vicinity of the compensation point small variations in stoichiometry of the Fe–Tb nanodots have large impact on their net magnetization and thus strong influence on the SFD.

References

1. J.-G. Zhu, Y. Tang, J. Appl. Phy. **99**, 08Q903 (2006)
2. D. Suess, J. Fidler, K. Porath, T. Schrefl, D. Weller, J. Appl. Phy. **99**, 08G905 (2006)
3. J.-G. Zhu, Y. Tang, IEEE Trans. Magn. **43**, 687 (2007)
4. L. Néel, Annuales de Géographica **5**, 99 (1949)
5. H. Kronmuller, M. Durst, J. Magn. Magn. Mater. **74**, 291 (1988)
6. C. Schulze, M. Faustini, J. Lee, H. Schletter, M. Lutz, P. Krone, M. Gass, K. Sader, A. Bleloch, M. Hietschold, Nanotechnology **21**, 495701 (2010)
7. M. Grobis, C. Schulze, M. Faustini, D. Grosso, O. Hellwig, D. Makarov, M. Albrecht, Appl. Phy. Lett. **98**, 192504 (2011)
8. Y. Mimura, N. Imamura, T. Kobayashi, A. Okada, Y. Kushiro, J. Appl. Phy. **49**, 1208 (1978)
9. B. Lanchava, H. Hoffmann, J. Phys. D Appl. Phys. **31**, 1991 (1998)
10. E. Stoner, E. Wohlfarth, Phil. Trans. R. Soc. Lond. A **240**, 599 (1948)
11. E. Kondorsky, J. Phy. (Moscow) **2**, 161 (1940)

Chapter 8
Interfacial Exchange Coupling in Heterostructures of Fe–Tb Alloy Films and Co/Pt Multilayers

The magnetic exchange between RE and TM elements, which is based on the interaction between electrons of the 4f–5d RE orbitals and the 3d TM orbitals, allows strong interfacial exchange coupling in heterostructures consisting of an amorphous FI RE–TM alloy film and a F TM film as discussed in Chap. 4. To elucidate this, exchange-biased heterostructures consisting of amorphous FI Fe–Tb alloy films and F Co/Pt multilayers were investigated. In particular, the dependence of the interfacial exchange coupling on the stoichiometry and thickness of the Fe–Tb layer as well as the number of repetitions in the Co/Pt multilayers were analyzed. Beside structural investigations and the measurement of integral magnetic properties element specific XMCD absorption experiments yield information about the magnetic sublattice configurations within different external fields and temperatures. This allows a detailed interpretation of the reversal processes and the dependency of the EB field present in the exchange coupled heterostructures. Furthermore, the training effect as a function of the temperature, the cooling field, and the magnitude of the cycling field was investigated in a heterostructure consisting of a Tb dominated Fe–Tb alloy film and a Co/Pt multilayer.

A series of samples with varying atomic stoichiometry of the Fe–Tb alloy film or different bilayer number of the Co/Pt multilayers were fabricated via magnetron (co-)sputtering using standard Si(100) wafers with a 100-nm-thick SiO_2 layer and 200-nm-thick Si_3N_4 membranes as substrates. The heterostructures are embedded in Pt seed and capping layers, which serve as oxidation barriers. Furthermore, the Pt seed layers provide good conditions for the growth of Co/Pt multilayers with PMA. Detailed information about the structural and magnetic properties in particular the reversal processes with respect to the interfacial exchange coupling will be given below.

C. Schubert, *Magnetic Order and Coupling Phenomena*, Springer Theses,
DOI: 10.1007/978-3-319-07106-0_8,

Fig. 8.1 Transmission electron microscopy images taken at two different magnifications in cross-section geometry of a $Fe_{81}Tb_{19}$(20 nm)/[Co(0.4 nm)/Pt(0.8 nm)]$_{10}$ heterostructure embedded in Pt seed and capping layers revealing some roughness and intermixing at the Fe–Tb/[Co/Pt] interface [1]

8.1 Morphology and Structural Properties

Transmission electron microscopy imaging and electron diffraction revealed that all Fe–Tb films in the heterostructures exhibit an amorphous structure in the composition range from 14 to 30 at.% of Tb similar to single films as discussed in Sect. 6.1. Cross-section micrographs used to study the morphology of the interface between the two magnetic layers in the different sets of samples reveal interfacial roughness and intermixing within the range of 1–2 nm, as seen in the TEM image taken from a $Fe_{81}Tb_{19}$(20 nm)/[Co(0.4 nm)/Pt(0.8 nm)]$_{10}$ heterostructure (Fig. 8.1).

8.2 Interfacial Exchange Coupling with Regard to the Dominant Moment in the Fe–Tb Alloy System

Similar to the intrinsic behavior of the Fe–Tb alloy system (see Sects. 2.2 and 6.2), the interfacial exchange interaction between Fe–Tb films and Co/Pt multilayers manifests itself in a parallel alignment of the Co and Fe moments and an antiparallel alignment of the Co and Tb moments (see Fig. 8.2) [2]. Varying the stoichiometry of the Fe–Tb alloy films across the magnetization compensation point, the dominant exchange coupling at the interface will change from F to AF. This is expected to influence the magnetization reversal of the whole heterostructure. To elucidate this, a series of Pt(3 nm)/$Fe_{100-x}Tb_x$(20 nm)/[Co(0.4 nm)/Pt(0.8 nm)]$_{10}$/Pt(4.2 nm)/substrate samples with a Tb content x varying from 14 to 30 at.% was prepared.

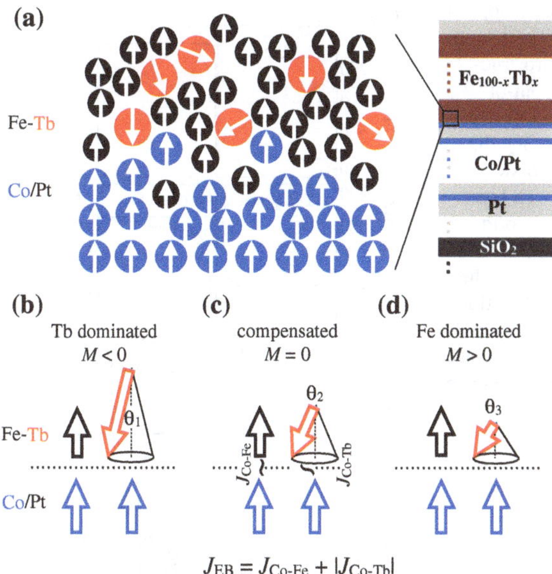

Fig. 8.2 **a** General schematic view of the magnetic configuration in heterostructures with PMA consisting of FI amorphous Fe–Tb alloy films and F Co/Pt multilayers. The *black* and the *blue arrows* represent the collinear alignment of the Fe and Co sublattices, respectively. In addition a drawing of a sample layer stack is given on the *right*. The *red arrows* with different length and opening angles θ illustrate the magnetic moment of the Tb sublattices and their average orientational distribution depending on the composition. **b–d** Average magnetic configuration illustrated by macro-spins depending on the composition of the Fe–Tb alloy film. The opening angles $\theta_{1...3}$ of the Tb macro-spins refer to the different fanning cone structures of the Tb sublattice. **b** The net magnetization of the Fe–Tb alloy film is dominated by the Tb sublattice. **c** The magnetic sublattices of the Fe and the Tb compensate each other and the net magnetization is zero. **d** The net magnetization of the Fe–Tb alloy film is dominated by the Fe sublattice. Please note that the different states depend on the Tb content and temperature

8.2.1 Interfacial Exchange Coupling in Fe Dominated Heterostructures

In Fig. 8.3a the magnetization reversal of $Fe_{81}Tb_{19}$(20 nm)/[Co(0.4 nm)/Pt(0.8 nm)]$_{10}$ heterostructures obtained from SQUID measurements along the out-of-plane direction at different temperatures is shown. The reversal behavior of this bilayer system is dominated by a F exchange coupling between Fe and Co at the interface. Around RT both layers reverse simultaneously in a single process. With decreasing temperature the switching field of the Fe–Tb layer increases much stronger as compared to the Co/Pt multilayers. This leads to a two-step reversal of the magnetization (Fig. 8.3a e.g., at 70 and 140 K). The separate switching fields of both layers allows the unidirectional anisotropy to be revealed through a shift of the hysteresis loop of the Co/Pt multilayers. In our experiments, this occurs as soon as the switching field of the

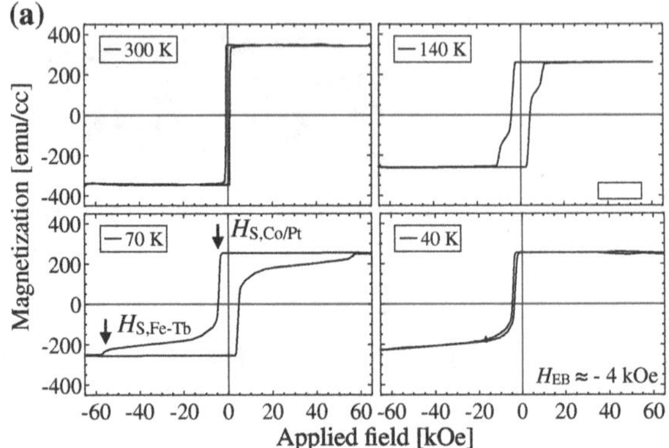

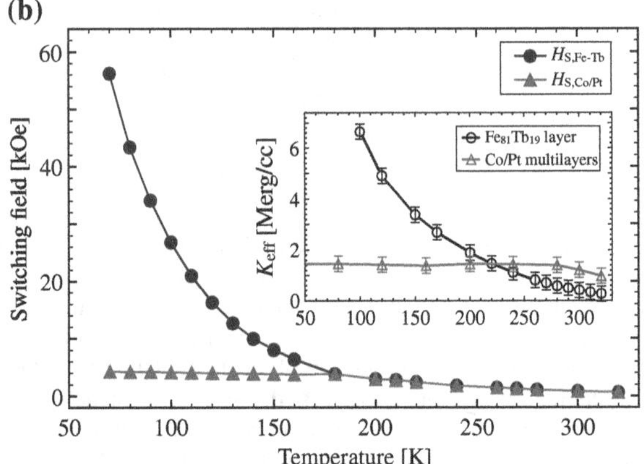

Fig. 8.3 **a** Hysteresis loops of $Fe_{81}Tb_{19}$(20 nm)/[Co(0.4 nm)/Pt(0.8 nm)]$_{10}$ heterostructures measured in out-of-plane geometry at different temperatures using SQUID magnetometry and **b** switching field of the Fe–Tb layer ($H_{S,Fe-Tb}$) and the Co/Pt multilayer ($H_{S,Co/Pt}$) taken from the *right* branch of the hysteresis loop as a function of the temperature compared to the temperature dependence of K_{eff} of the single reference layers (inset)

Fe–Tb layer becomes larger than the maximum applied external field. In particular, using a maximum external field of 70 kOe, the unidirectional anisotropy sets in at a temperature of 40 K, leading to a loop shift with an EB field of $H_{EB} \approx -4$ kOe.

An overview of the switching fields of the Fe–Tb layer ($H_{S,Fe-Tb}$) and the Co/Pt multilayer ($H_{S,Co/Pt}$) as a function of temperature as compared to K_{eff} of the single reference layers is given in Fig. 8.3b. Please note that the switching field was estimated from the derivative of the magnetization curve in the reversal region using the peak center of a Gaussian fit function. In the temperature range from RT down to about 180 K the anisotropies of both layers are more or less the same leading to the observed simultaneous switching (Fig. 8.3a e.g., 300 K). Furthermore, the sharp switching of the hysteresis loops indicate a magnetization reversal driven by domain wall nucleation and propagation, which is possible due to the strong exchange coupling of the Co and Fe sublattices at the interface. The region below 180 K is characterized by a strong enhancement of the Fe–Tb switching field having its origin in the increase of the magnetic anisotropy and the decrease of the net magnetization with decreasing temperature (see Fig. 6.7a in Sect. 6.3).

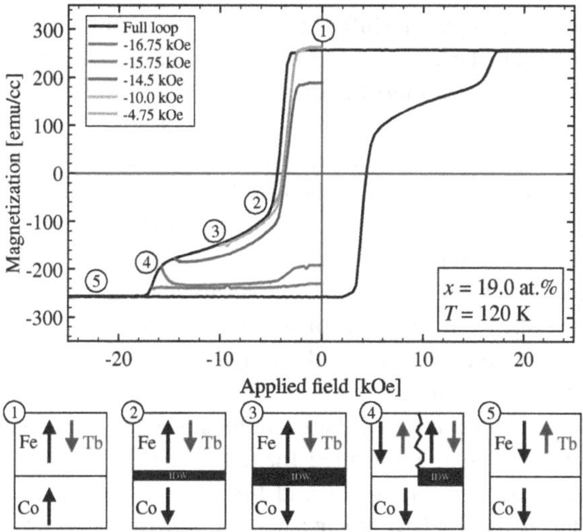

Fig. 8.4 Minor loop reversal taken from out-of-plane SQUID measurements and a schematic drawing of the assumed domain configuration during the magnetization reversal in $Fe_{81}Tb_{19}$(20 nm)/[Co(0.4 nm)/Pt(0.8 nm)]$_{10}$ heterostructures

In exchange coupled heterostructures the reversal of one layer against the other will produce an IDW (see Chap. 4) in order to minimize the cost of energy for the magnetic transition region between both layers. Referring to the present system, with increasing external magnetic field the switching of the Co/Pt multilayers against the Fe–Tb alloy film occurs together with the formation of an IDW since in its ground state the Co and Fe sublattices prefer a parallel alignment, while the Tb sublattice favors pointing antiparallel to them. If now the external field will be reduced coming from this AF configuration, at a certain field the IDW vanishes forcing the Co/Pt multilayers to rotate back so that the system can relax in its ground state. Due to the strong interfacial exchange coupling, the high amount of energy stored in the IDW gives rise to the large loop shift observed in this system.

The minor loop reversal study presented in Fig. 8.4 provides potential evidence for the occurrence of an IDW in the AF configuration of the Fe–Tb alloy film and the Co/Pt multilayer. Starting from the remanent state after saturation, where the dominant moments Fe and Co are aligned parallel as illustrated by the scheme number 1, an out-of-plane external field starts switching the Co/Pt multilayer leading to the formation of an IDW (scheme number 2). This behavior is completely reversible since the magnetization ends up in the same remanent state after reducing the field to zero. Even though the reversal field is increased further, as long as the switching field of the Fe–Tb layer is not reached, the reversal mechanism does not change. With regard to this, one can assume that the reversible characteristic is driven by nucleation and annihilation of the IDW during the field cycling. Furthermore, the reversibility in the higher field range can be attributed to an increasing or decreasing wall thickness depending on the applied field (see scheme number 3). Going on from the cartoon 3 to 4 in the hysteresis loop, the increasing field starts to overcome the switching field of the Fe–Tb layer. In this intermediate state, where the Fe–Tb is not completely reversed, a partial shift of the minor loops still exists. This indicates that the reversal of the Fe–Tb layer is driven by the nucleation and expansion of lateral domains. These domains appear most likely as bubbles or stripes, which will not vanish in zero-field. This produces the configuration shown in the cartoon 4 of Fig. 8.4, in which some part of the heterostructure still possesses an IDW giving rise to the partially shifted minor loops observed, when the field is reduced to zero. In particular, the minor loop produced by a reversal field of −15.75 kOe shows an unusual behavior, since the magnetization drops even more after reducing the external field. This might have its origin in a peculiar domain expansion during the reversal of the Fe–Tb layer. Finally, a further enhancement of the external field ends up in a complete reversal of the Fe–Tb layer modifying the remanent state (see cartoon number 5) since the IDW vanishes at saturation.

A confirmation of these findings is given by element specific XMCD absorption measurements performed for the $Fe_{83}Tb_{17}$(20 nm)/[Co(0.4 nm)/Pt(0.8 nm)]$_{10}$ heterostructure. The minor and full hysteresis loops of the Co, Fe, and Tb moments during the field cycling at a temperature of 160 K are shown in Fig. 8.5. From the full hysteresis loops one can find that the sign of the normalized XMCD signal changes for the Tb moments as compared to the Fe and Co moments referring to the antiparallel alignment between the rare-earth and transition-metal sublattices [3]. In addition, a small shoulder occurs in the hysteresis loops of the Fe and Tb moments during the reversal of the Co moments and vice versa. These features can be attributed to the proposed formation of an IDW, because the nucleation of such a domain wall will reduce the XMCD signal, which is proportional to the out-of-plane component of the magnetic moment. The portion of the shoulder as compared to the whole signal amounts to about 14 % for Co, 11 % for Fe, and 7 % for Tb, which can be used to estimate the element specific portion of the domain wall width t_{IDW}. Please note that the difference in the percentages between the Tb and Fe signals is caused most likely by the non-collinearity (fanning cone structure) of the Tb moments, contributing less to the decrease in the signal by the nucleation of the IDW. Taking into account the Co and Fe percentages together with the thickness of the Co/Pt multilayers ($t_{Co/Pt} = 12$ nm) and the Fe–Tb layer ($t_{Fe-Tb} = 20$ nm) one can estimate the total domain wall width to $t_{IDW} = t_{IDW,Co/Pt} + t_{IDW,Fe-Tb} = (3.8 \pm 0.3)$ nm. Assuming that the IDW exhibits a structure similar to an ultra-thin Néel wall, it is possible to determine the effective exchange stiffness A_{eff} of the material in the interface region using the following equation [4]:

$$A_{eff} = \frac{t_{IDW}^2}{4} \cdot K_{eff}. \tag{8.1}$$

With an effective anisotropy calculated from the values $K_{eff,Co/Pt} = (1.4 \pm 0.3)$ Merg/cm^3 for the Co/Pt multilayers and $K_{eff,Fe-Tb} = (2.3 \pm 0.4)$ Merg/cm^3 for the Fe–Tb alloy film as obtained from the reference samples at 160 K weighted by the thickness of the IDW ranging into each layer the effective stiffness at the interface results in $A_{eff} = (6.8 \pm 2.6) \times 10^{-8}$ erg/cm. In comparison to literature values [5, 6], for Fe–Tb alloy films (1×10^{-7} erg/cm) and Co/Pt multilayers (5×10^{-7} erg/cm)

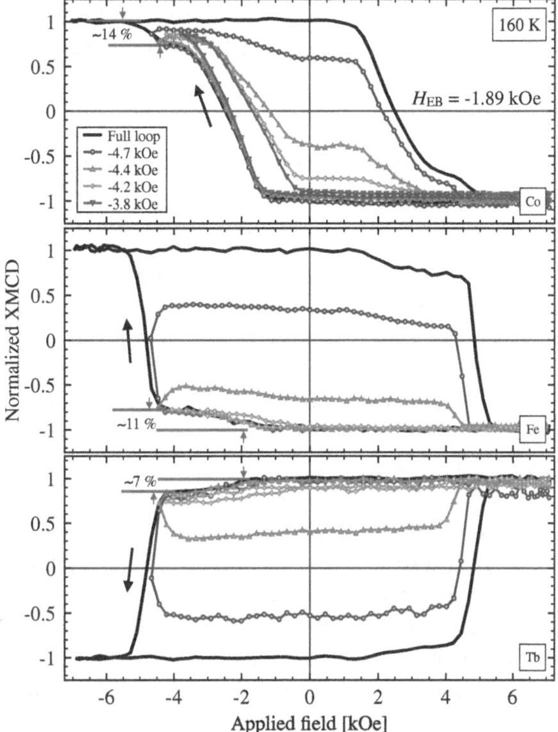

Fig. 8.5 Element specific hysteresis loops of $Fe_{83}Tb_{17}$(20 nm)/[Co(0.4 nm)/Pt(0.8 nm)]$_{10}$ heterostructures obtained from XMCD absorption measurements at the L_3-edge of Fe (708 eV) and Co (778 eV) and the M_5-edge of Tb (1241 eV) at 160 K with minor and full loop field cycling. The percentages indicate the portion of the XMCD signal related to the IDW

the effective exchange stiffness is reduced at the interface. A reason for this can be the formation of a mixed layer at the Fe–Tb/[Co/Pt] interface.

Based on this result one can recalculate the EB field via the energy density of the IDW, σ_{IDW}, using Eq. 4.2 according to Sect. 4 and the following relation: [4, 7–10]

$$\sigma_{IDW} = 2\pi\sqrt{A_{eff} \cdot K_{eff}} = \pi \cdot t_{IDW} \cdot K_{eff}. \tag{8.2}$$

With regard to the estimated domain wall energy density of $\sigma_{IDW} = (2.2 \pm 0.7)$ erg/cm^2, by using the measured values for t_{IDW} and K_{eff}, the EB field results to $H_{EB} = -(1.4 \pm 0.6)$ kOe. Please note that the specified error values were assessed by assuming a reduced anisotropy value at the interface according to the observed effective exchange stiffness. In comparison to the measured EB field of -1.89 kOe (see Fig. 8.5) the value estimated by the IDW model is in good agreement. For the calculations the magnetostatic interaction was not taken into account since only a marginal effect on the EB field due to the reduced magnetization of the Fe–Tb layer, is expected, which will be discussed in more detail later. Calculations using the model of Meiklejohn and Bean [11] with the assumption of a closed packed structure of the atoms in the Fe–Tb alloy film lead to an almost 10 times higher value. This can be seen as the maximum possible EB field for the material system appearing if the system would not develop an IDW in order to minimize the total energy.

8.2.2 Interfacial Exchange Coupling in Tb Dominated Heterostructures

With regard to the AF exchange coupling between rare-earth and transition-metal moments, the magnetic configuration of the dominant moments in heterostructures consisting of Fe–Tb films and Co/Pt multilayers changes from F to AF with increasing amount of Tb in the amorphous alloy film. This behavior is due to the fact that the net magnetization in amorphous Fe–Tb alloy films becomes dominated by the Tb moment when the Tb content is greater than 24 at.% (Fig. 6.7a). In this paragraph we will present an investigation on the interfacial exchange coupling in heterostructures exhibiting an AF configuration of the dominant moments. As a representative example, we will focus on the magnetization reversal behavior of the $Fe_{75.5}Tb_{24.5}$(20 nm)/[Co(0.4 nm)/Pt(0.8 nm)]$_{10}$ heterostructure.

The out-of-plane hysteresis loops in Fig. 8.6a give an overview of the magnetization reversal of the chosen bilayer system at different temperatures. In contrast to the simultaneous reversal in the Fe dominated heterostructures, as described in the prior section, one obtains a separate switching of the magnetic layers already at RT. Thereby, the Fe–Tb layer manifests its reversal in the appearance of satellite hysteresis loops at higher magnetic fields. This behavior can be attributed to a dominant AF exchange coupling between the two layers and the nucleation of an IDW formed if the magnetization of both layers is aligned parallel to each other. With increasing temperature (e.g., 390 K) the observed satellite hysteresis loops do not vanish but become smaller, which is associated to the lowering of the saturation magnetization in the vicinity of T_C. Concerning the magnetization reversal of the Co/Pt multilayers one observes an enhanced coercive field. In contrast to this behavior at lower temperatures (e.g., 90 K) the heterostructure relaxes after saturation in a positive field into its AF remanent state by reversing the magnetization of the Co/Pt multilayers instead of the Fe–Tb film. This leads to the onset of a positive EB field $H_{EB,Co/Pt} \approx +4$ kOe of the Co/Pt multilayers, if the switching field of the Fe–Tb layer becomes larger than the external field (e.g., 10 K). A similar observation was made for in-plane F/AF and FI/FI systems with AF interface coupling [12, 13].

An explanation for the observed reversal characteristic of the heterostructure can be deduced from the temperature dependence of the switching fields of the Fe–Tb layer and the Co/Pt multilayers compared to their effective magnetic anisotropy obtained from the single reference layers (inset of Fig. 8.6b). As long as the temperature stays above 230 K the anisotropy of the Fe–Tb layer remains smaller than the Co/Pt multilayers. Furthermore, the saturation magnetization and with it the Zeeman energy of the Fe–Tb alloy film is much smaller as compared to the Co/Pt multilayers. Therefore, the reversal back into the preferred AF configuration under the

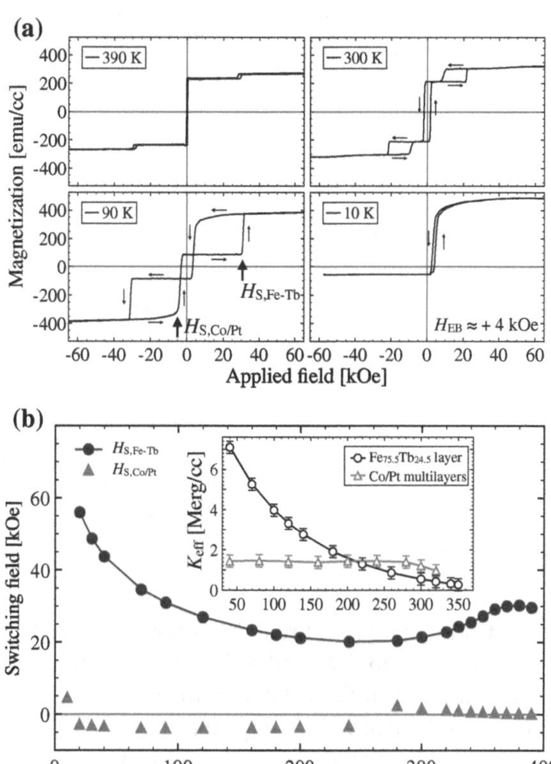

Fig. 8.6 **a** Hysteresis loops of $Fe_{75.5}Tb_{24.5}$(20 nm)/[Co(0.4 nm)/Pt(0.8 nm)]$_{10}$ heterostructures measured in out-of-plane geometry at different temperatures using a SQUID magnetometer and **b** switching field of the Fe–Tb layer ($H_{S,Fe-Tb}$) and the Co/Pt multilayers ($H_{S,Co/Pt}$) as a function of temperature compared to the temperature dependence of K_{eff} of the single reference layers (inset). Please note that the increase in the switching field towards higher temperatures is related to the compensation point of the Fe–Tb layer, which is close to the Curie temperature

presence of an external field costs less energy for the Fe–Tb layer than for the Co/Pt multilayers. However, below 260 K the situation changes, and the Co/Pt multilayers reverse against the Fe–Tb layer towards remanence (Fig. 8.6a, 90 K). The reason for this behavior lies in the Fe–Tb layer gaining magnetic anisotropy (inset of Fig. 8.6b) and net magnetization in the lower temperature regime. Thus, the energy barrier for the magnetization reversal consisting of the anisotropy and Zeeman energy, becomes larger for the alloy film as compared to the Co/Pt multilayers favoring the observed switching. Finally, at temperatures below 20 K the switching field of the Fe–Tb layer due to the enhanced magnetic anisotropy overcomes the maximum applied magnetic field. This results in the observed positive EB field of the Co/Pt multilayers.

A detailed understanding of the detected AF interaction between the Fe–Tb alloy film and the Co/Pt multilayers via the formation of an IDW can be derived from the minor loop study at 120 K presented in Fig. 8.7. Scheme number 1 refers to the saturation configuration in the positive field direction at about 30 kOe. In this state the dominant sublattice moments Co and Tb aligned parallel, and a domain wall owing to the minimization of the AF exchange energy characterizes the interface region. Coming from this point, the reduction of the external field gives rise to the reversal

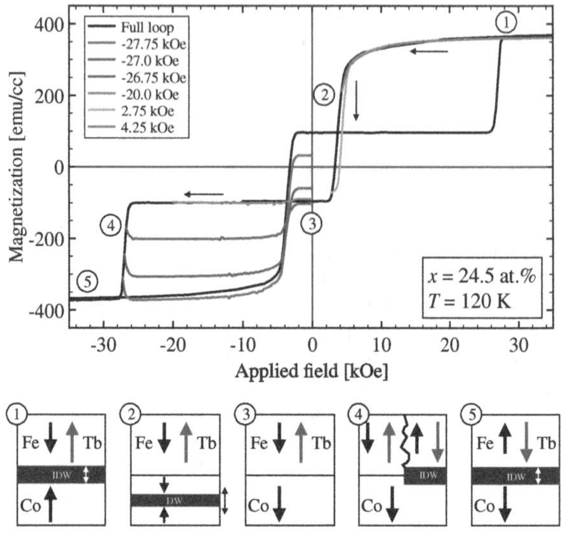

Fig. 8.7 Minor loop reversal taken from out-of-plane SQUID measurements and a schematic drawing of the assumed domain configuration during the magnetization reversal in $Fe_{75.5}Tb_{24.5}$(20 nm)/ [Co(0.4 nm)/Pt(0.8 nm)]$_{10}$ heterostructures

of the Co/Pt multilayer. From the minor loops produced by the field cycling starting from saturation to 2.75 and 4.25 kOe one finds a completely reversible behavior. This indicates that the extinction of the IDW might involve the propagation of a transversal domain wall through the Co/Pt multilayers (see scheme number 2). The antiparallel alignment of the dominant moments, displayed in scheme number 3, illustrates the assumed zero-field configuration of the heterostructure, which is free from any domain wall state. In contrast to the Co/Pt multilayer, the Fe–Tb alloy film shows non-reversibility in the magnetization during the negative field cycling. This can be attributed to the nucleation of lateral domains (see scheme number 4). These domains are highly stable and produce a domain wall at the interface to the corresponding part of the Co/Pt multilayer leading to their partial switching observed when the field is reduced to zero. In a saturation field of about −30 kOe the heterostructure reveals the same magnetic configuration as in positive field direction (scheme number 5).

In comparison to heterostructures with a preferred F coupling, it was found that the interfacial exchange coupling whether F or AF manifests itself mainly in the sign change of the EB field. Nevertheless, a slight influence of the Fe–Tb layers composition on the magnitude of the exchange coupling was found and will be discussed in the following paragraph.

8.2.3 The Interfacial Exchange Energy as Function of the Composition of the Fe–Tb Layer

The EB field represents a measure of the interfacial exchange energy density, J_{EB}, based on the relation given in Eq. 4.2. Figure 8.8a, b summarize J_{EB} and the magnitude of the EB field, respectively, as a function of the Tb content in $Fe_{100-x}Tb_x$/[Co/Pt]$_{10}$ heterostructures at 10 and 120 K. Furthermore, the total EB field is outlined in the inset indicating the sign change from minus to plus at the compensation point when the dominant sublattice moment changes from Fe to Tb. At 10 and 120 K a similar composition dependency becomes apparent, only the total values differ slightly. This is attributed to the increasing effective magnetic anisotropy of the Fe–Tb film towards lower temperatures, which leads to an increasing IDW energy (see Eq. 8.2) and a high value for J_{EB}. From 18 to 25 at.% Tb the exchange energy density seems to reach its maximal value. Towards lower and higher amounts of Tb some reduction appears. This behavior can not be explained by the variation of the effective magnetic anisotropy of the Fe–Tb alloy films (see Fig. 6.9a in Sect. 6.3). Despite its strong relation to the IDW energy and therefore to the interfacial exchange energy (see Eq. 4.2 in Sect. 4 and Eq. 8.2), the changes in K_{eff} are far too small to explain the observed composition dependency completely. Instead the sperimagnetic spin configuration and in consequence the net magnetization of the Fe–Tb alloy film might play also an important role. In the recent work of Radu et al. [14] as well as Ungureanu et al. [15] the authors find, that the EB field reaches its maximum at the compensation point, since the compensated sublattices of the FI film hold no frustrated bonds at the interface to the F layer. With regard to this the net magnetization has an impact on the exchange energy. However, in our case the influence is rather small as observed from the composition dependency of the net magnetization shown in Fig. 6.9b of Sect. 6.3. While the magnetization increases from almost zero at 21 at.% to 400 emu/cm^3 towards lower and higher amounts of Tb the exchange energy per unit area decreases only by about 20 %.

With regard to the maximum value of $J_{EB} = 3.5\,\text{erg/cm}^2$ at 10 K the $Fe_{100-x}Tb_x$/[Co/Pt]$_{10}$ heterostructures exhibit by a factor of 10 higher exchange energy densities as other comparable perpendicular exchange-bias thin film systems such as [Co/Pt]/CoO heterostructures [16].

8.3 Interfacial Exchange Coupling Depending on the Thickness of the F Layer

The magnitude of the EB field in an F/AF heterostructure with varying thickness of the F layer follows typically an inverse relation [9, 10]. However, this is only valid for high anisotropic AF materials, which yield a certain amount of uncompensated frozen spins at the interface and disallow the nucleation of a domain wall in the layer itself. In contrast to this, the EB effect in the present F/FI heterostructure is based

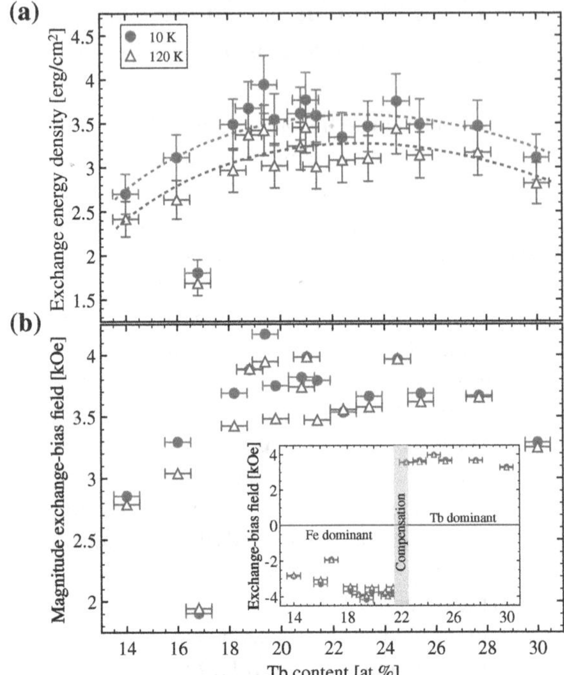

Fig. 8.8 **a** The interfacial exchange energy density and **b** the magnitude of the EB field as a function of the Tb content obtained from $Fe_{100-x}Tb_x$/[Co/Pt]$_{10}$ heterostructures at 10 and 120 K. The dashed curves act as a guide to the eye. The error for the EB field lies within the size of the symbols. And the scattering of data points is most likely related to the interface roughness induced by changes in growth conditions depending on the Tb content of the alloy. Please note that the sign of the EB field changes at a composition of about 22 at.% due to the transition of the dominant magnetic moment from Fe to Tb in the Fe–Tb alloy film as shown in the inset

on the nucleation of an IDW storing the exchange energy. Therefore one can expect a different dependency of the EB field on the F layer thickness as compared to the F/AF system.

To elucidate this expectation, a series of $Fe_{80.5}Tb_{19.5}$(20 nm)/[Co(0.4 nm)/Pt(0.8 nm)]$_n$ heterostructures was analyzed with regard to the magnitude of the EB field as function of the [Co/Pt] bilayer number n by varying n from 4 to 16. The thickness of the Fe–Tb alloy film was held constant at 20 nm.

From SQUID and XMCD absorption measurements a nearly linear decrease of the magnitude of the exchange-bias field with increasing thickness of the F layer was found (see Fig. 8.9a). Except for a small variation in the magnitude this dependence does not change with temperature. An understanding of the linear behavior provides the layer specific thickness of the IDW as function of the Co/Pt multilayer thickness as presented in Fig. 8.9b. While the total thickness of the IDW stays more less the same, interestingly their location changes with increasing F layer thickness. This manifests itself in the simultaneous reduction and enhancement of the IDWs parts located in the Co/Pt multilayers and in the Fe–Tb layer, respectively. The effective magnetic anisotropy of the Co/Pt multilayers, as obtained from reference samples, changes by about 20% with the number of repetitions from 1.7 Merg/cm^3 for $n = 4$ to 2.2 Merg/cm^3 for $n = 10$. An increase of the number of repetitions to $n = 16$ yield no further anisotropy enhancement. Therefore, the magnetic anisotropy can not explain the observed shift completely. Another reason might be the gain in

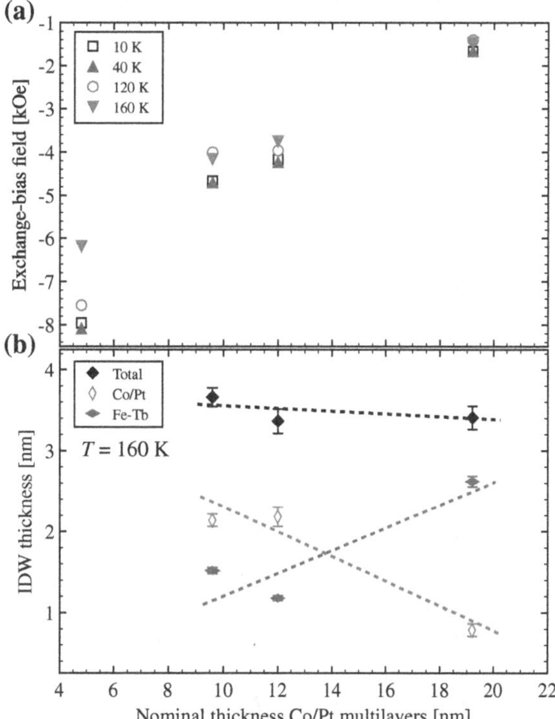

Fig. 8.9 **a** Exchange-bias field $H_{EB,Co/Pt}$ of the Co/Pt multilayers at 10, 40, 120, and 160 K and **b** the total as well as partial IDW thickness ($t_{IDW} = t_{IDW,Co/Pt} + t_{IDW,Fe-Tb}$) related to both layers of the heterostructure at 160 K drawn as a function of the nominal thickness of the multilayer stack obtained from SQUID and XMCD absorption measurements. The *dashed lines* act as a guide to the eye

magnetic moment towards higher thicknesses increasing the Zeeman energy. In consequence the domain wall moves further into the Fe–Tb layer, until the equilibrium with respect to the magnetic anisotropy energy and exchange energy of the FI material is reached. The Zeeman energy, which follows linearly the magnetic moment, might lead to a corresponding linear shift. Since the Fe–Tb alloy film yields a smaller domain wall energy density, σ_{DW}, (3.5 erg/cm^2) as compared to the Co/Pt multilayers (5.2 erg/cm^2) due to the different values for the exchange stiffness [5, 6] the change in the position of the IDW might cause the linear reduction of the EB field. A change of the location where the IDW resides is also known from other rare-earth-transition-metal-based heterostructures [13, 17, 18].

8.4 Interfacial Exchange Coupling Depending on the Thickness of the FI Pinning Layer

The EB effect, as based on the interfacial exchange coupling between magnetic bilayers, is often considered as interface effect depending on the density of frozen uncompensated spins of the AF or FI pinning layer in the transient region to the adjacent

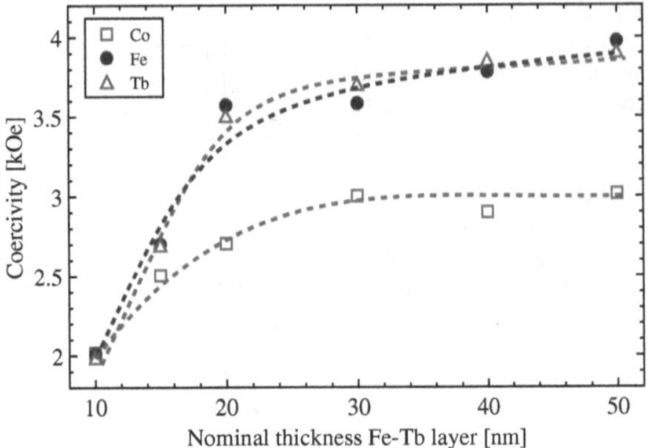

Fig. 8.10 Element specific coercivity as a function of the nominal Fe–Tb layer thickness in $Fe_{80}Tb_{20}(t_{Fe-Tb})$/[Co/Pt]$_{10}$ heterostructures obtained from field dependent XMCD absorption measurements at RT. The *dashed lines* act as a guide to the eye

F layer [19–21]. Nevertheless, the magnetic moments in the volume of the pinning layer can also contribute to the EB effect according to investigations presented in the literature [22–26]. Particularly, in case of Fe–Tb based EB heterostructures, where the nucleation of an IDW with finite size carries the uniaxial anisotropy, the intrinsic magnetic properties in the volume of the FI pinning layer influence the EB effect. In order to elucidate the influence of the pinning layer volume exchange coupled heterostructures consisting of $Fe_{80}Tb_{20}$ films with varying thickness t_{Fe-Tb} from 10 to 50 nm and [Co(0.4 nm)/Pt(0.8 nm)]$_{10}$ multilayers were fabricated on 200-nm-thick Si_3N_4 membranes using magnetron (co-)sputtering and analyzed by field dependent XMCD absorption measurements at different temperatures.

The coercivity as a function of the Fe–Tb layer thickness for $Fe_{80}Tb_{20}$/[Co/Pt] heterostructures is shown in Fig. 8.10. The values are obtained separately for the three elements Co, Fe, and Tb from element specific hysteresis loops at RT. Heterostructures with a 10-nm-thick Fe–Tb layer possess a coercive field of about 2 kOe with a simultaneous reversal of the Co/Pt multilayer and the Fe–Tb film. With increasing thickness the coercivity of the Fe–Tb film becomes enhanced up to 4 kOe. Due to the strong interfacial, exchange coupling the coercivity of the Co/Pt multilayer increases as well, although it follows not completely. The reversal in these heterostructures occurs in two steps via separate switching of the Co/Pt multilayer and the Fe–Tb film. Generally, the coercivity enhancement is attributed to an increase in effective magnetic anisotropy for thicker Fe–Tb films as discussed in Sect. 2.3.

The two step magnetization reversal in heterostructures with Fe–Tb layer thicknesses above 15 nm allows the onset of an unidirectional loop shift for the Co/Pt multilayer within a minor loop, where the magnetic alignment of the Fe–Tb film remains unchanged. The magnitude of the resulting EB field for the Co/Pt multilayers is presented in Fig. 8.11 as a function of the Fe–Tb layer thickness and the temperature. As found previously the EB field increases slightly towards lower temperatures due to the increasing PMA of the Fe–Tb film. Independently from the temperature the EB field becomes also enhanced with increasing Fe–Tb layer thickness for the same reason. Since the magnetic anisotropy is correlated with the energy

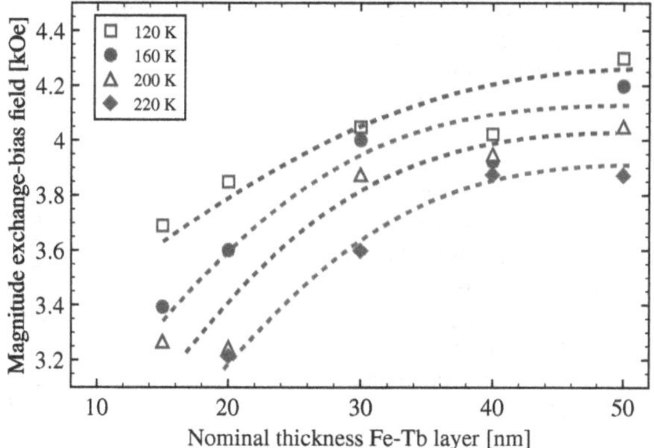

Fig. 8.11 Magnitude of the exchange-bias field as a function of the nominal Fe–Tb layer thickness in $Fe_{80}Tb_{20}(t_{Fe-Tb})$/[Co/Pt]$_{10}$ heterostructures obtained from field dependent XMCD absorption measurements at the L_3-edge of Co for various temperatures. The *dashed lines* act as a guide to the eye

density of the IDW according to Eq. 8.2 the gain in PMA due to the increasing Fe–Tb layer thickness causes the enhancement of the EB field. Based on this result the EB effect in this particular system is also related to the volume properties of the FI pinning layer, since the conditions of the Fe–Tb/[Co/Pt] interface were kept constant and only the pinning layer thickness was increased. However, a detailed understanding of the mechanism between the magnetic anisotropy of the volume material and the EB field is missing and requires further investigations.

8.5 Training Effect in EB Heterostructures Dominated by an AF Exchange

One of the main characteristics in artificial EB bilayer structures is the training effect describing the difference between subsequent magnetization hysteresis loops. This effect was primary reported by Paccard et al. in 1966 [27]. Since then a change in the EB field depending on the number of field cycles referred to as training was found in F/AF, F/FI, and FI/FI based exchange coupled heterostructures [13, 28–35]. Generally the training effect manifests itself in a reduction of the EB field between subsequent field cycles, which can be described empirically following a reciprocal square root relation or a recursive formula [28, 29].

Contrary to this common observation exchange coupled heterostructures consisting of a Co/Pt multilayer stack and a Fe–Tb alloy film reveal a remarkable increase, denoted as ΔH_{EB}, of the EB field between the first and the second field cycle, while additional field cycles yield no further changes. This can be seen in the progress of the EB field depending on the applied field cycle of a $Fe_{75}Tb_{25}$(20 nm)/[Co(0.4 nm)/Pt(0.8 nm)]$_{10}$ heterostructure at 10 K presented in Fig. 8.12. The measurement of the EB field training was realized by reversing the magnetization of the Co/Pt multilayer against the Fe–Tb film within a minor loop field cycling after cooling the sample in an applied field of 50 kOe from RT.

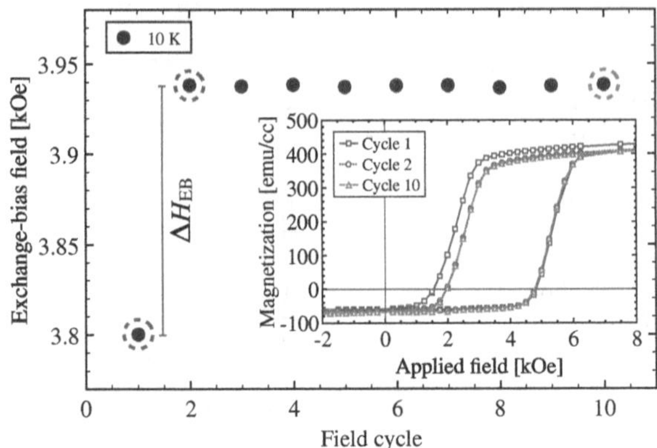

Fig. 8.12 Exchange-bias field as a function of the magnetic field cycling of a $Fe_{75}Tb_{25}/[Co/Pt]_{10}$ heterostructure obtained from minor hysteresis loop measurements at 10 K using a SQUID magnetometer. The difference in the EB field from the first to the second cycle is denoted with ΔH_{EB}. The inset reveals the minor hysteresis loops of the first, second and tenth field cycle

As the Fe–Tb alloy film possesses a Tb sublattice moment dominated net magnetization, the parallel alignment of the Tb and Co moments in the high field range causes the nucleation of an IDW due to the inappropriate energetic state (see Sect. 8.2.2). In this manner the moment configuration of the IDW becomes cooled down and determines the reversal of the Co/Pt multilayer within at least the first field cycle at the measurement temperature. The change of the hysteresis loop from the first to the second field cycle is shown in the inset of Fig. 8.12. With reducing the external applied field for the first time after field cooling the reversal of the Co/Pt multilayer against the Fe–Tb film into the preferred antiparallel configuration takes place at lower fields as compared to the second time. This results in the observed enhancement of the EB field by ΔH_{EB}. An explanation for this training effect can be deduced from the characteristic of the EB effect, which is determined by the properties of the IDW in this particular system. As discussed in Sect. 8.2 the value of the EB field correlates with the energy of the IDW and thus with the effective magnetic anisotropy as well as exchange stiffness at the interface region of the adjacent magnetic layers. Due to the strong temperature dependence of these magnetic properties in the Fe–Tb alloy system (see Sect. 6.3), the energy of the IDW exhibits an enhancement towards lower temperatures. Assuming that the spin configuration remains unchanged during field cooling, the reversal of the Co/Pt multilayer into the AF configuration within the first field cycle is dominated by the configuration and energy of the IDW nucleated at RT. In further field cycles the reversal of the Co/Pt multilayer yield an IDW, which nucleates and vanishes always with the same properties with regard to the measurement temperature and provides no additional changes in H_{EB}. Therefore, the enhancement of the EB field from the first to the second field cycle results from the higher domain wall energy in the low temperature regime compared to RT.

Please note, the higher saturation magnetization in the hysteresis loop of the first field cycle compared to the second one (see the inset in Fig. 8.12) might be correlated to the moment configuration of the IDW. However, the occurrence of this effect is not pronounced and a clear understanding requires further element specific XMCD

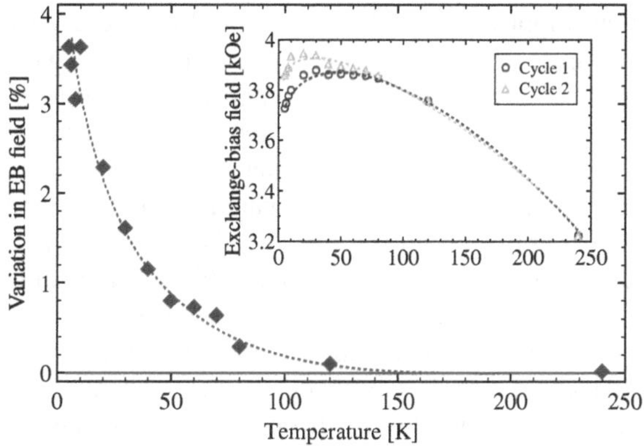

Fig. 8.13 Temperature dependence of the variation of the EB field from the first to the second field cycle of a $Fe_{75}Tb_{25}/[Co/Pt]_{10}$ heterostructure. The EB field of the first and second field cycle as a function of the temperature is presented in the inset. The *dashed curves* provide a guide to the eye. All values were received from magnetization hysteresis loops measured at several temperatures after cooling from RT within a magnetic field of 50 kOe

absorption measurements in order to obtain information of the thickness of the IDW as a function of the field cycling.

8.5.1 Characteristic of the Training Effect with Respect to the Temperature

Previously the general characteristic of the EB training effect in Fe–Tb/[Co/Pt] heterostructures was discussed to originate from the temperature dependence of the IDW energy. In order to gain a comprehensive understanding of the effect, the variation of the EB field between the first and second field cycle ΔH_{EB} was investigated as a function of temperature. The variation ΔH_{EB} normalized to the EB field value of the first cycle is shown in Fig. 8.13. The inset reveals the progress of the EB field of the first and second field cycle with respect to the temperature.

Towards lower temperatures starting from 250 K the EB field becomes enhanced by about 20 % due to the increasing effective PMA and exchange stiffness of the Fe–Tb layer providing a higher IDW energy (see Sect. 8.2). Below 10 K the EB field decreases slightly again, which might be attributed to a shift of the IDW location into the Co/Pt multilayer stack due to the strong increase in PMA in the Fe–Tb layer. However, a complete understanding of the origin of this behavior can not be provided at the moment as it requires a detailed investigation of the orientational moment configuration within the domain wall at the interface.

At about 80 K H_{EB} of the first and second field cycle start to differ and a remarkable training effect sets in. The variation of the EB field increases with decreasing temperature, which originates from the increasing difference between the energy of the IDW at RT and at the measurement temperature. With regard to Eq. 8.2 and the temperature dependence of the effective magnetic anisotropy in a 20-nm-thick $Fe_{75}Tb_{25}$ single layer (see Fig. 6.8 in Sect. 6.3) the anisotropy enhancement of more than a factor of 7 below 80 K produces more than 2.5 times higher energy values for the IDW in comparison to the RT value. Compared to the variation in the EB field with the temperature a similar progress can be found although the value remains only in the range up to 4 % (see Fig. 8.13). A reason for this deviation might be related to the temperature dependence of the exchange stiffness as well as the location, where the domain wall becomes nucleated at the interface. Since the magnetic properties of the Co/Pt multilayer stack are hardly changing with temperature it is possible that the IDW shifts into the Co/Pt multilayer with decreasing temperature in order to minimize its energy consumption due to the gain in magnetic anisotropy of the Fe–Tb film. Nevertheless, clear evidence for this assumption is still missing and requires a detailed analysis of the thickness and location of the IDW as a function of the field cycling and temperature, which will be realized perspectively via element specific XMCD absorption measurements.

8.5.2 Influence of the Cooling Field on the Training Effect

A cooling field dependence for the training of the EB field is observed in an exchange coupled $Fe_{75}Tb_{25}/[Co/Pt]_{10}$ heterostructure in the range of up to 1.5% as exemplarily outlined in Fig. 8.14 for a temperature of 40 K. The inset shows the behavior of the EB field of the first and second field cycle as a function of the cooling field. A very striking point within this dependency is the zero field cooling, which leads to a maximum value of H_{EB} and a vanishing training effect. This is explainable with the assumption that a more or less collinear interfacial spin structure becomes frozen due to the cooling of the heterostructure in its appropriate AF configuration between the Fe–Tb film and the Co/Pt multilayer. In this manner, no domain wall and canted spin structure exists, which avoids frustration at the interface. Thus, the IDW nucleates first time at the measurement temperature leading to similar reversal characteristics in the Co/Pt multilayer for the first, second, and all other field cycles. Furthermore, the maximum EB field originates most likely also from the frozen collinear interfacial spin structure, which forces the IDW to nucleate with small thickness and under high energy consumption.

For a cooling field different from zero the characteristic changes in so far as a smaller EB field and a variation between the first subsequent field cycles are observed. Towards higher applied cooling fields the normalized variation ΔH_{EB} decreases from about 1.5% at 10 kOe to about 1% at 70 kOe, while the magnitude of H_{EB} increases. The reason for this particular behavior lies in the manner how the IDW becomes nucleated at RT and afterwards frozen to the measurement temperature. A small cooling field like 10 kOe almost saturates the magnetic moments of the Co/Pt multilayer parallel to the Tb dominated net moment of the Fe–Tb layer, but it

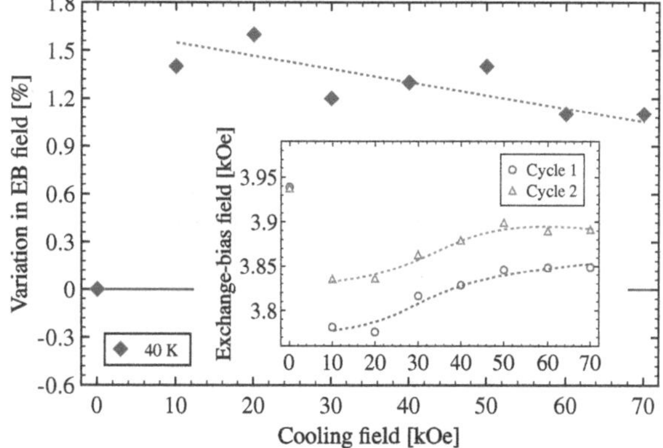

Fig. 8.14 The variation of the EB field from the first to the second field cycle as a function of the applied cooling field of a $Fe_{75}Tb_{25}$/[Co/Pt]$_{10}$ heterostructure at 40 K. The corresponding EB field values of the first and second field cycle are shown in the inset. The *dashed curves* act as a guide to the eye

also allows the nucleation of the IDW with rather broad canted moment configuration at RT. Therefore, the difference of the properties of the IDW nucleated during the second field cycle at the measurement temperature becomes significantly larger and the frozen canted interfacial spins provide a smaller EB field value compared to the collinear configuration resulting from zero field cooling. With higher cooling fields the RT IDW exhibits a smaller canted moment distribution giving rise for a smaller training effect and higher H_{EB} as more collinear spins at the interface become frozen during field cooling.

8.5.3 Cycle Field Dependence of the EB Training Effect

Besides the temperature and the cooling field another parameter yielding influence on the training effect of the EB field is given by the cycle field, denoted as the maximum positive field applied during the minor loop cycling. In Fig. 8.15 the cycle field dependence of the normalized EB field variation ΔH_{EB} for a $Fe_{75}Tb_{25}$/[Co/Pt]$_{10}$ heterostructure at 10 K is presented. The inset reveals the progress of H_{EB} of the first and second field cycle. While the absolute values of the EB field possess only a small variation with the cycle field, ΔH_{EB} decreases by 0.85% from 3.45 to 2.6% by a change of the cycle field from 10 to 70 kOe, respectively. In particular the reduction occurs mainly for cycle fields up to 50 kOe. For higher fields a saturation sets in, which is obviously correlated with the applied cooling field of 50 kOe. If one assumes that the spin configuration of the IDW, which is nucleated at RT becomes partially frozen during the field cooling process, the canted frozen spin configuration yield an influence on the nucleation and vanishing of the IDW during the minor loop magnetization reversal of the Co/Pt multilayer. In consequence the frozen spins contribute to the EB effect. Since such a frozen spin configuration is only fixed up to a certain value of the external applied field, higher cycle fields lead to smaller EB fields as well as a smaller training effect. The saturation of this behavior for cycling

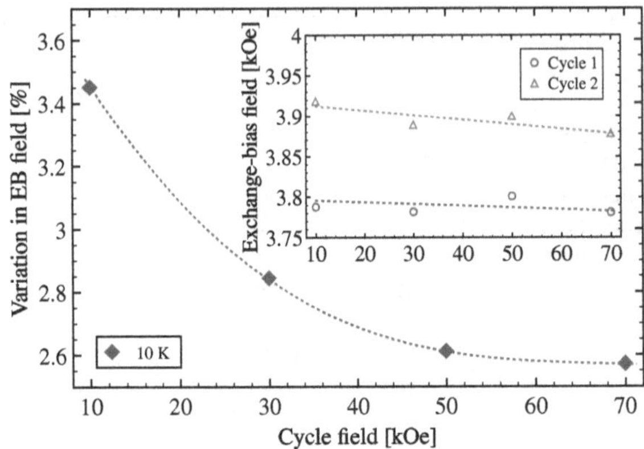

Fig. 8.15 Cycle field dependence on the variation of the EB field from the first to the second field cycle of a $Fe_{75}Tb_{25}/[Co/Pt]_{10}$ heterostructure at 10 K. The EB fields of the first and second field cycle, which are presented in the inset, were measured from the magnetization hysteresis loops in an applied magnetic field varying between −10 kOe and the several positive cycle fields after cooling in a field of 50 kOe from RT. Please note that for each cycle field a new cooling procedure was applied. The *dashed curves* serve as a guide to the eye

fields above the cooling field results due to the fact that the spin configuration is no longer influenced by frozen spins, which allows the nucleation of an IDW with equilibrium spin configuration according to the interfacial magnetic properties and applied external field. However, a frozen interfacial spin configuration might be only one possible explanation for the observed behavior and further investigations are necessary to provide clarity as mentioned in the previous paragraphs.

8.6 Summary

In conclusion, the interfacial exchange coupling in perpendicular heterostructures consisting of FI amorphous Fe–Tb alloys and F Co/Pt multilayers was investigated. From element specific hysteresis loops obtained by XMCD absorption measurements, an IDW with a finite size of 3–4 nm was found to exist, when the Co moments are forced to point antiparallel to the Fe moments and parallel to the Tb moments. Due to the exchange interaction between rare-earth and transition-metal moments, this state is energetically inappropriate and the system tries to minimize the energy by nucleating an IDW.

The magnitude of the exchange energy per unit area depends on the net magnetization and effective magnetic anisotropy of the Fe–Tb layer, which was probed by varying the amount of Tb in the film. It was found that with reducing net magnetization towards zero at 21 at.% Tb the exchange energy reaches its maximum. Furthermore, at lower temperatures a small increase was observed, which is attributed to the increasing effective magnetic anisotropy of the Fe–Tb film.

Although in common EB systems an inverse relation between the exchange field and the thickness of the F layer is expected, the investigated heterostructures show a more linear dependency, which can be attributed to the IDW moving linearly into the Fe–Tb layer with increasing thickness of the Co/Pt multilayers. This leads to a maximum EB field of about 8 kOe in heterostructures with a nominal Co/Pt multilayer thickness of 4.2 nm.

Concerning the question wether the EB effect is related solely to the interface or also to the volume of the pinning layer, the EB field as a function of the FI Fe–Tb pinning layer thickness was investigated. The increasing thickness of the Fe–Tb layer results in a gain of PMA for the alloy film, which yield significant influence on the EB field. Although the morphology of the interface region is identical, the increasing thickness of the Fe–Tb pinning layer leads to higher EB fields due to the magnetic anisotropy. Therefore, the volume of the pinning layer is crucial for the unidirectional anisotropy in this particular system.

Finally the training of the EB effect by subsequent field cycling of a Fe–Tb/[Co/Pt] heterostructure with dominant AF interfacial exchange was analyzed. In contrast to the common observations for the training effect the system reveals an increasing EB field from the first to the second field cycle and no further changes occur in additional field cycles. The enhancement of the EB field during training is explainable with a difference in spin configuration and energy consumption for the IDW nucleated at RT and at the measurement temperature, which determine independently the reversal of the Co/Pt multilayer stack during the first and second field cycle, respectively. Furthermore, the spin configuration within the IDW might be partially frozen during field cooling providing also an impact on the training effect especially with respect to the cooling and cycle field dependence.

The presented results give rise for a phenomenological understanding of the exchange field characteristic, however an appropriate quantitative model is still missing at the moment. Thus, further investigations are necessary to understand the EB effect in this particular system.

References

1. H. Schletter, Präparation und Charakterisierung nanostrukturierter Magnetwerkstoffe unter besonderer Berücksichtigung des Exchange Bias Effekts, Ph.D. Thesis, TU Chemnitz (2013)
2. S. Mangin, T. Hauet, P. Fischer, D. Kim, J.B. Kortright, K. Chesnel, E. Arenholz, E.E. Fullerton, Phys. Rev. B **78**, 024424 (2008)
3. I. Campbell, J. Phys. F Met. Phys. **2**, L47 (1972)
4. S. Middelhoek, J. Appl. Phys. **34**, 1054 (1963)
5. Y. Mimura, N. Imamura, T. Kobayashi, A. Okada, Y. Kushiro, J. Appl. Phys. **49**, 1208 (1978)
6. S. Hashimoto, Y. Ochiai, J. Magn. Magn. Mater. **88**, 211 (1990)
7. T. Kobayashi, H. Tsuji, S. Tsunashima, S. Uchiyama, Jpn. J. Appl. Phys. **20**, 2089 (1981)
8. S. Tsunashima, J. Phys. D Appl. Phys. **34**, R87 (2001)
9. J. Nogués, I. Schuller, J. Magn. Magn. Mater. **192**, 203 (1999)
10. R. Stamps, J. Phys. D Appl. Phys. **33**, R247 (2000)
11. W.H. Meiklejohn, C.P. Bean, Phys. Rev. **102**, 1413 (1956)

12. J. Nogués, C. Leighton, I. Schuller, Phys. Rev. B **61**, 1315 (2000)
13. T. Hauet, J. Borchers, S. Mangin, Y. Henry, S. Mangin, Phys. Rev. Lett. **96**, 067207 (2006)
14. F. Radu, R. Abrudan, I. Radu, D. Schmitz, H. Zabel, Nat. Commun. **3**, 715 (2012)
15. M. Ungureanu, K. Dumesnil, C. Dufour, N. Gonzalez, F. Wilhelm, A. Smekhova, A. Rogalev, Phys. Rev. B **82**, 174421 (2010)
16. S. Maat, K. Takano, S. Parkin, E.E. Fullerton, Phys. Rev. Lett. **87**, 87202 (2001)
17. T. Hauet, S. Mangin, F. Montaigne, J.A. Borchers, Y. Henry, Appl. Phys. Lett. **91**, 022505 (2007)
18. S.M. Watson, T. Hauet, J.A. Borchers, S. Mangin, E.E. Fullerton, Appl. Phys. Lett. **92**, 202507 (2008)
19. S. Roy, M. Fitzsimmons, S. Park, M. Dorn, O. Petracic, I. Roshchin, Z. Li, X. Batlle, R. Morales, A. Misra et al., Phys. Rev. Lett. **95**, 47201 (2005)
20. F. Radu, H. Zabel, Springer Tracts Mod. Phys. **227**, 97 (2008)
21. K. O'Grady, L. Fernandez-Outon, G. Vallejo-Fernandez, J. Magn. Magn. Mater. **322**, 883 (2009)
22. T. Ambrose, C.L. Chien, J. Appl. Phys. **83**, 6822 (1998)
23. H. Sang, Y.W. Du, C.L. Chien, J. Appl. Phys. **85**, 4931 (1999)
24. S.M. Zhou, K. Liu, C.L. Chien, J. Appl. Phys. **87**, 6659 (2000)
25. M. Lund, W. Macedo, K. Liu, J. Nogués, I. Schuller, C. Leighton, Phys. Rev. B **66**, 054422 (2002)
26. M. Ali, C. Marrows, M. Al-Jawad, B. Hickey, A. Misra, U. Nowak, K. Usadel, Phys. Rev. B **68**, 214420 (2003)
27. D. Paccard, C. Schlenker, O. Massenet, R. Montmory, A. Yelon, Phys. Status Solidi **16**, 301 (1966)
28. A. Hoffmann, Phys. Rev. Lett. **93**, 097203 (2004)
29. C. Binek, Phys. Rev. B **70**, 014421 (2004)
30. I. Guhr, O. Hellwig, C. Brombacher, M. Albrecht, Phys. Rev. B **76**, 064434 (2007)
31. T. Hauet, S. Mangin, J. Mccord, F. Montaigne, E.E. Fullerton, Phys. Rev. B **76**, 144423 (2007)
32. S. Sahoo, S. Polisetty, C. Binek, A. Berger, J. Appl. Phys. **101**, 053902 (2007)
33. S. Sahoo, T. Mukherjee, K.D. Belashchenko, C. Binek, Appl. Phys. Lett. **91**, 172506 (2007)
34. M. Patra, M. Thakur, S. Majumdar, S. Giri, J. Phys. D Appl. Phys. **21**, 236004 (2009)
35. S. Giri, M. Patra, S. Majumdar, J. Phys. D Appl. Phys. **23**, 073201 (2011)

Chapter 9
Interlayer Exchange Coupling Through Pt Spacer Layers in Fe–Tb/Pt/[Co/Pt] Heterostructures

The interfacial exchange coupling between FI amorphous Fe–Tb alloy films and F Co/Pt multilayers leads to the formation of an IDW in the inappropriate magnetic configuration, where Co and Tb moments are pointing parallel to each other. This IDW gives rise for the appearance of the EB effect as discussed in the Chap. 8. In this manner, the interface in particular its morphology plays a crucial role for the formation of the IDW and finally for the onset of the EB effect. In order to explore the influence of the interface morphology and coupling constant on the EB field, heterostructures with different Pt spacer layers were fabricated. The series of heterostructures consist of 20-nm-thick $Fe_{81}Tb_{19}$ films and $[Co(0.4\,nm)/Pt(0.8\,nm)]_{10}$ multilayers separated by Pt layers with thicknesses ranging from 0 nm to 5 nm, respectively. Standard Si(100) wafers with a 100-nm-thick SiO_2 layer and 200-nm-thick Si_3N_4 membranes serve as substrates for the film deposition.

In the following a high resolution TEM investigation in cross-section geometry probing the interface region of $Fe_{81}Tb_{19}(20\,nm)/Pt(t_{Pt})/[Co(0.4\,nm)/Pt(0.8\,nm)]_{10}$ heterostructures with $t_{Pt} = 0 \ldots 5$ nm will be presented. This provides detailed information about the interface morphology depending on the Pt spacer layer thickness. Please note, the structural TEM investigations were performed by H. Schletter within a project cooperation and the results are briefly presented according to his Ph.D. thesis [1]. Based on the structural results a model will be discussed to explain the dependence of the EB field on the thickness of the Pt spacer layer.

9.1 Morphology and Structural Properties

From the cross-section TEM investigation of $Fe_{100-x}Tb_x/[Co/Pt]_{10}$ heterostructures without Pt spacer layer, as presented in Sect. 8.1, an interfacial roughness and intermixing within the range of 1–2 nm was observed. Thus, the interface region possesses an effective exchange stiffness and anisotropy constant, which determines the configuration of the IDW and in consequence the EB effect. With an additional Pt spacer

C. Schubert, *Magnetic Order and Coupling Phenomena*, Springer Theses,
DOI: 10.1007/978-3-319-07106-0_9,

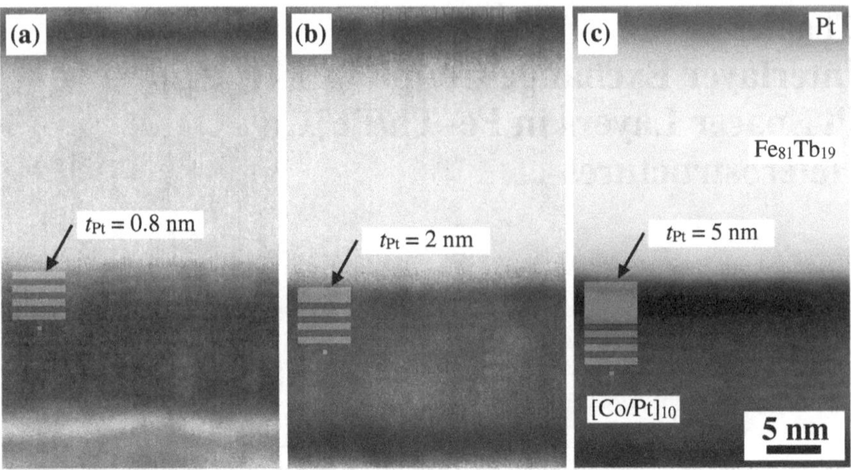

Fig. 9.1 High resolution bright field TEM images in cross-section geometry of $Fe_{81}Tb_{19}$(20 nm)/Pt(t_{Pt})/[Co(0.4 nm)/Pt(0.8 nm)]$_{10}$ heterostructures with different Pt interlayer thicknesses of **a** 0.8 nm, **b** 2 nm, and **c** 5 nm, respectively [1]. The alternating blue and grey lines indicate the subsequent layers of the Co/Pt multilayer stacks

layer the interface morphology becomes modified and changes in magnetic properties particularly the exchange coupling are to be expected leading to variations in the EB field. In order to analyze the interface morphology with respect to the Pt spacer layer thickness a comprehensive high resolution bright field TEM study was performed. Images of $Fe_{81}Tb_{19}$(20 nm)/Pt(t_{Pt})/[Co(0.4 nm)/Pt(0.8 nm)]$_{10}$ heterostructures with spacer layer thicknesses t_{Pt} of 0.8, 2, and 5 nm are given in Fig. 9.1a, b, c, respectively.

Independent from the Pt spacer layer thickness the $Fe_{81}Tb_{19}$ films grow in an amorphous structure with similar magnetic properties, as observed in preliminary studies (not shown). All heterostructures reveal a partially layered structure, which is attributed to the Co/Pt multilayers. An illustration of the subsequent Co and Pt layers is provided by the blue and grey lines in the TEM images of Fig. 9.1, respectively. The Co/Pt multilayers grow polycrystalline with (111) texture, which leads to local differences in the interface roughness due to the different oriented crystallites. Some interface regions are flat, while others exhibit large notches attributed to step edges and grain boundaries of several crystallites. This can be seen as the origin for the observed interface roughness and intermixing between the Fe–Tb film and the Co/Pt multilayer in the heterostructures.

According to the structure and morphology of the Co/Pt multilayer one can assume that samples with an additional Pt spacer layer with thicknesses up to 5 nm reveal similar structures. In particular the Pt spacer layer with a thickness of 0.8 nm follows closely the morphology of the multilayer stack as observed in the TEM image of Fig. 9.1a. Consequently, spacer layers with thicknesses less than 1 nm are not closed and with regard to the notches between crystallites full exchange coupling between the Fe–Tb film and the Co/Pt multilayer can occur locally. Thicker Pt spacer layers

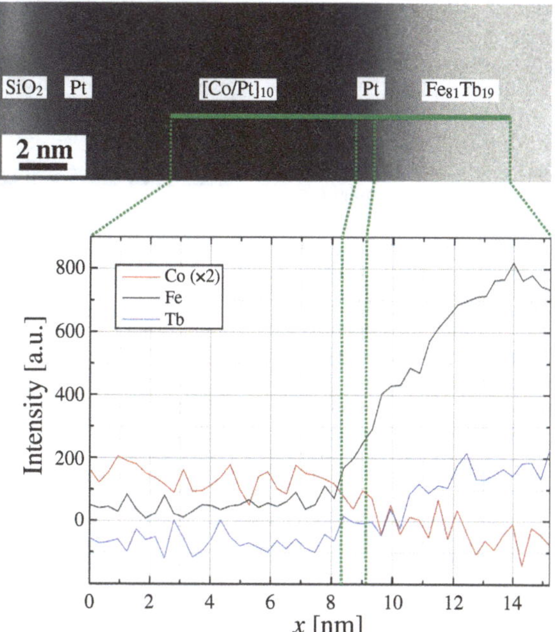

Fig. 9.2 High resolution bright field TEM image in cross-section geometry of a $Fe_{81}Tb_{19}$(20 nm)/Pt(0.8 nm)/[Co(0.4 nm)/Pt(0.8 nm)]$_{10}$ heterostructure with EELS line profile obtained from the interface region highlighted in *green* [1]. The signal intensity coming from the Co L-edge was multiplied by a factor of two for better representation. Please note that the negative values in the spectra originate from the normalization and yield no physical relevance

seem to be continuous, although they follow also closely the morphology of the Co/Pt multilayers (see Fig. 9.1b). However, their thickness is large enough to cover the notches originating from next neighbor crystallites with different orientations of the Co/Pt multilayers. Thus, magnetic exchange coupling appears not due to a direct contact between the Fe–Tb films and the Co/Pt multilayers, nevertheless an indirect exchange is possible provided by spin polarization present in the Pt spacer layer [2, 3].

In order to analyze the degree of intermixing at the interface, the local distribution of Co, Fe, and Tb was measured using electron energy-loss spectroscopy (EELS) [4]. The EELS line profiles of $Fe_{81}Tb_{19}$(20 nm)/Pt(t_{Pt})/[Co(0.4 nm)/Pt(0.8 nm)]$_{10}$ heterostructures with Pt spacer layer thicknesses of 0.8 and 5 nm are presented together with the corresponding high resolution cross-section TEM images in Figs. 9.2 and 9.3, respectively.

The heterostructure with 0.8-nm-thick Pt spacer layer reveals a significant overlap of the Co and Fe signals in the EELS profile of the interface region. This is attributed to a possible intermixing, as the thin spacer layer provides no sufficient intermixing barrier. Furthermore, the notches of neighboring Co/Pt multilayer crystallites, which are filled with amorphous Fe–Tb, will also contribute to the Co and Fe EELS signals. In contrast to this, the EELS line profile within the interface region in Fig. 9.3 shows a clear separation of the Co and Fe signals for the heterostructures with 5-nm-thick spacer layer. Thus, intermixing between the Co/Pt multilayer and the Fe–Tb film can be excluded. Based on the TEM investigation with EELS of heterostructures with

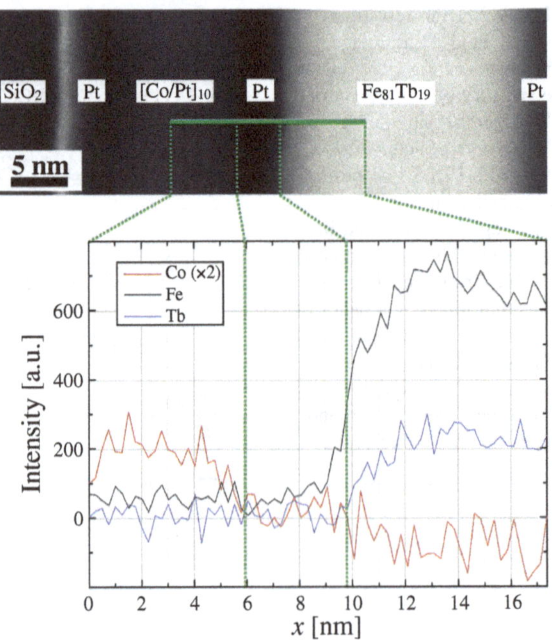

Fig. 9.3 High resolution bright field TEM image in cross-section geometry of a $Fe_{81}Tb_{19}$(20 nm)/ Pt(5 nm)/[Co(0.4 nm)/ Pt(0.8 nm)]$_{10}$ heterostructure with EELS line profile obtained from the interface region highlighted in *green* [1]. The signal intensity coming from the Co L-edge was multiplied by a factor of two for better representation. Please note that the negative values in the spectra originate from the normalization and yield no physical relevance

thinner Pt layers (not shown) a sufficient intermixing barrier is provided for spacer layers down to a thickness of about 2 nm.

Concerning the impact of the notches between neighboring Co/Pt multilayer crystallites on the morphology of the Pt spacer layer and the Fe–Tb film, the interface region was investigated in detail. A high resolution bright field TEM image in cross-section geometry of a $Fe_{81}Tb_{19}$(20 nm)/Pt(0.8 nm)/[Co(0.4 nm)/Pt(0.8 nm)]$_{10}$ heterostructure in the region of two neighboring Co/Pt multilayer crystallites is presented in Fig. 9.4a. As mentioned before, the 0.8-nm-thick Pt spacer layer follows the morphology of the Co/Pt multilayer crystallites. The white dashed curves in the TEM image indicate the course of the interface in the vicinity of the notch. Please note, the weak contrast in the image originates from overlapping additional crystallites due to the prepared TEM sample thickness of more than 10 nm and the observed lateral crystallite size of below 10 nm (see Sect. 6.1). At the point of contact between the two Co/Pt multilayer crystallites the Pt spacer layer covers the last Co layer not completely. Therefore, amorphous Fe–Tb films is in direct contact to the Co/Pt multilayer, which provides local magnetic exchange coupling. A structural model of the interface is outlined in Fig. 9.4b. In general, the size of the notches remains in the range of 1–2 nm leading to the assumption that interfacial intermixing and direct magnetic exchange coupling between the Co/Pt multilayer and the Fe–Tb alloy film can occur up to a spacer layer thickness of 2 nm. For more details about the structural analysis of the heterostructures with Pt spacer layer please pay attention to the Ph.D. thesis of Schletter [1].

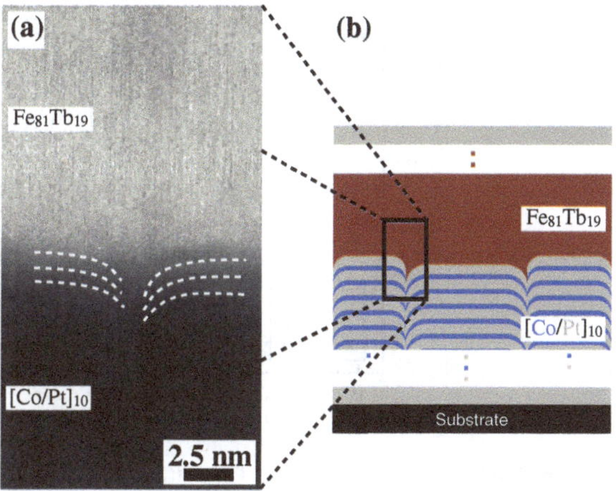

Fig. 9.4 **a** High resolution bright field TEM image in cross-section geometry of a $Fe_{81}Tb_{19}$(20 nm)/Pt(0.8 nm)/[Co(0.4 nm)/Pt(0.8 nm)]$_{10}$ heterostructure [1]. The *dashed bright curves* indicate the layer structure of the Co/Pt multilayer stack. **b** Schematic drawing of the morphology of the Co/Pt multilayers at the interface to the Pt interlayer and the Fe–Tb film

9.2 Influence of Pt Spacer Layer Thickness on the EB Field

With regard to the interface morphology the structural investigations revealed that the interlayer exchange coupling in $Fe_{81}Tb_{19}$/Pt(t_{Pt})/[Co/Pt]$_{10}$ heterostructures can be attributed to a direct magnetic exchange between the Co/Pt multilayers and the Fe–Tb films at the notches between neighboring crystallites of the Co/Pt multilayers supplemented by an indirect exchange based on the spin polarization in the Pt spacer layers. To elucidate the exchange coupling, the EB field was measured as a function of the nominal Pt spacer layer thickness after the samples were saturated at RT and cooled in zero field to 10 K. The EB field normalized to the value for a sample without Pt spacer layer is presented in Fig. 9.5. The inset reveals the shifted magnetization hysteresis loops of the Co/Pt multilayers for different spacer layer thicknesses at 10 K. Please note, for the data obtained from SQUID hysteresis loops the samples were saturated in a magnetic field of 70 kOe and for element specific XMCD hysteresis loops measured at the Co L_3-edge a saturation field of 7 kOe was applied. Despite the different RT saturation fields similar values for the EB fields were found, which is a consequence of the zero field cooling in remanence. Otherwise differences depending on the cooling field are to be expected, as discussed in Sect. 8.5.2.

Heterostructures with Pt spacer layers up to a thickness of 0.8 nm reveal more or less no changes in the value of the EB field. Even 2-nm-thick spacer layers allow 50 % of the initial field value. For larger spacer layers the exchange coupling becomes strongly decreased as far as a 4-nm-thick Pt layer suppresses the EB effect completely. To understand this behavior one need to separate two effects.

Firstly, up to a thickness of 0.8 nm the Pt layer is not closed and local direct exchange between the Co/Pt multilayers and the Fe–Tb alloy films is possible, as deduced from the interface morphology presented in the previous section. Due to this and the spin polarization of the Pt, the IDW can nucleate in the whole interface region and provide the maximum EB field. In particular, from XMCD absorption

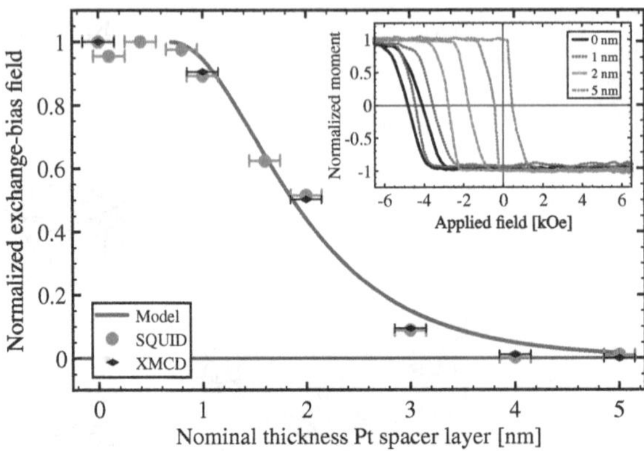

Fig. 9.5 Exchange-bias field of $Fe_{81}Tb_{19}$/Pt(t_{Pt})/$[Co/Pt]_{10}$ heterostructures at 10 K as a function of the nominal Pt spacer layer thickness t_{Pt} normalized to the EB field value for the heterostructure without Pt spacer. The *red line* represents the fit of the data to the corresponding model. The inset reveals the magnetization hysteresis loops of the heterostructures with different Pt spacer layer thickness. Please note, within a cooperation these results were provided to Schletter and preliminarily published in his Ph.D. thesis [1]

measurements a IDW thickness of about 3 nm was found in heterostructures with 0.8-nm-thick spacer layer.

Secondly, spacer layers with a thickness larger than 0.8 nm are more or less closed and the interlayer exchange occurs most likely via spin polarization of the Pt. In this manner, both interfaces (i.e., the Co/Pt and the Pt/Fe–Tb interface) cause a spin polarization in the Pt with similar decay or exchange length. Generally, the spin polarization becomes reduced with $\exp[-z/\delta]$, where δ refers to the exchange length and z is the coordinate perpendicular to the film plane. Taking into account the spin polarization originating from both interfaces the exchange between the Co/Pt multilayer and the Fe–Tb film through the Pt spacer layer is given by the sum of two decay functions describing the spin polarization p_s as a function of the z-coordinate and layer thickness:

$$p_s = \frac{e^{-z/\delta} + e^{-(t_{Pt}-z)/\delta}}{1 + e^{t_{Pt}/\delta}}. \tag{9.1}$$

The minimum of this function is located at the half of the Pt layer thickness $t_{Pt}/2$ and gives the maximum exchange between the Co/Pt multilayers and the Fe–Tb alloy films and with it the normalized EB field depending on the spacer layer thickness:

$$\frac{H_{EB}}{H_{EB,0}} = 2 \cdot \frac{e^{-t_{Pt}/2\delta}}{1 + e^{t_{Pt}/\delta}}, \tag{9.2}$$

while $H_{EB,0}$ denotes the EB field for a heterostructure without Pt spacer layer. Please note, in order to obtain a good accordance between the experimental data and the model a Pt spacer layer thickness reduced by 0.8 nm is assumed, as thinner layers provide full exchange coupling. The data fit is shown together with the experimental data in Fig. 9.5. Based on the fit result an effective exchange length of $\delta = 0.4$ nm for the Pt spacer layer can be found.

9.3 Summary

The exchange coupling between amorphous FI Fe–Tb alloy films and F Co/Pt multilayers becomes strongly influenced by an additional Pt spacer layer. As the Co/Pt multilayers possess a polycrystalline structure with (111) texture the interface region is characterized by more or less flat crystallites. However, the flat regions are disconnected by 1–2 nm deep notches generated by neighboring crystallites. These notches allow most likely direct contact between the Co/Pt multilayer and the Fe–Tb film.

As observed from high resolution TEM imaging with EELS, the Pt spacer layer follows closely the morphology of the Co/Pt multilayer crystallites. Depending on the thickness of the Pt layer, the Co/Pt multilayers become more or less covered. Due to shadowing during film deposition at the notches between crystallites, Pt layers with thicknesses below 0.8 nm are not continuous. Thus, full magnetic exchange coupling between the Co/Pt multilayers and the Fe–Tb films occurs in heterostructures with Pt spacer layers up to a thickness of 0.8 nm leading to an EB field value similar to the heterostructure without spacer layer. Thicker Pt layers are continuous and cover the notches between neighboring Co/Pt multilayer crystallites, which allows only indirect magnetic exchange due to the spin polarization in the Pt. Consequently, the EB field becomes strongly reduced for spacer layers thicker than 0.8 nm. Based on the presented model taking into account the spin polarization in the Pt spacer layer, an exchange length of about 0.4 nm was found. With regard to the full exchange coupling up to a Pt spacer layer thickness of 0.8 nm an effective exchange length of 1.2 nm for the system Fe–Tb/Pt/$[Co/Pt]_{10}$ system can be specified.

References

1. H. Schletter, Präparation und Charakterisierung nanostrukturierter Magnetwerkstoffe unter besonderer Berücksichtigung des Exchange Bias Effekts, Ph.D. Thesis, TU Chemnitz (2013)
2. G.A. Bertero, R. Sinclair, C.H. Park, Z.X. Shen, J. Appl. Phy. **77**, 3953 (1995)
3. W. Weber, D. Wesner, D. Hartmann, G. Güntherodt, Phy. Rev. B **46**, 6199 (1992)
4. R. Egerton, *Electron Energy-Loss Spectroscopy in the Electron Microscope* (Springer, US, 2011). ISBN 978-1-4419-9583-4

Chapter 10
Conclusions

From the fundamental point of view amorphous sperimagnetic Fe–Tb alloy films in the form of single thin layers as well as within exchange coupled heterostructures yield interesting magnetic properties and peculiar magnetization reversal processes according to the non-collinear spin structure of the Fe and Tb sublattices. The sperimagnetic spin configuration, which arises from the growth induced structural anisotropy of the short range order in conjunction with the single ion anisotropy, influences noticeably the integral magnetic properties like net saturation magnetization, coercivity, and magnetic anisotropy. This was analyzed in thin amorphous Fe–Tb alloy films with varying stoichiometry. According to SQUID magnetometry and field dependent element specific XMCD absorption measurements an orientational moment distribution, designated as fanning cone, was found in the Fe and Tb sublattices of this films. The fanning cone structure gives rise for a strong temperature dependence of the saturation magnetization and PMA as, driven by the enhanced exchange coupling, the orientational moment distribution becomes narrow and the Tb sublattice magnetization as well as effective magnetic anisotropy increase towards lower temperatures. With this also the reversal process changes in particular for Fe dominated samples from stripe to bubble domain reversal. The strongest impact of the fanning cone structure on the magnetization reversal occurs in the vicinity of the compensation point, where the net magnetization vanishes. Then, a reversal of magnetic moments is only driven by a spin-flop transition in the high magnetic field regime (above 40 kOe) followed by an orientational relaxation of the magnetic moments into their equilibrium distribution with decreasing field.

Not only flat films possess peculiar magnetization reversal behavior, percolated amorphous Fe–Tb films fabricated by co-deposition on pre-patterned substrates yield different reversal processes. Although the Fe–Tb nanodots in the pre-patterned films are fully exchange coupled via the trench material, single domain magnetic states are observed. These single domain nanodots reveal a separate switching and act as pinning sites for the magnetic domains in the trench material. For the magnetization reversal of the trench material a depinning field of about 3 kOe was found. As far as the external applied field becomes larger depinning sets in and the domain walls

C. Schubert, *Magnetic Order and Coupling Phenomena*, Springer Theses,
DOI: 10.1007/978-3-319-07106-0_10, © Springer International Publishing Switzerland 2014

move between subsequent pinning sites providing the magnetization reversal. The switching of the single domain nanodots is a more or less nucleation-dominated reversal process as the angular dependence of the switching field follows a Stoner-Wohlfarth like behavior. Surprisingly, the net magnetization in this percolated FI Fe–Tb alloy system plays a minor role for the reversal behavior and pinning effects. However, in the vicinity of the compensation point small variations in stoichiometry of the nanodots have large impact on their net magnetization and thus strong influence on the SFD.

After the intrinsic magnetic configuration of thin amorphous FI Fe–Tb alloy films and nanostructures were investigated the interfacial exchange coupling between Fe–Tb films and adjacent Co/Pt multilayers became analyzed comprehensively. From element specific hysteresis loops obtained by XMCD absorption measurements, an IDW with a finite size of 3–4 nm was found to exist, when the Co moments are forced to point antiparallel to the Fe moments and parallel to the Tb moments. Due to the exchange interaction between rare-earth and transition-metal moments, this state is energetically inappropriate and the system tries to minimize the energy by nucleating an IDW. The magnitude of the exchange energy per unit area depends on the net magnetization and effective magnetic anisotropy of the Fe–Tb layer, which was probed by varying the amount of Tb in the film. It was found that with reducing net magnetization towards zero at 21 at.% Tb the exchange energy reaches its maximum. Although in common EB systems an inverse relation between the exchange field and the thickness of the F layer is expected, the investigated heterostructures show a more linear dependency, which can be attributed to the IDW moving linearly into the Fe–Tb layer with increasing thickness of the Co/Pt multilayers. This leads to a maximum EB field of about 8 kOe in heterostructures with a nominal Co/Pt multilayer thickness of 4.2 nm.

Furthermore, the interlayer exchange coupling though a Pt spacer layer was explored between Fe–Tb alloy films and Co/Pt multilayers in artificial heterostructures as a strong influence of the interface morphology on the EB effect was expected. According to this study full magnetic exchange coupling between Co/Pt multilayers and Fe–Tb films was found in heterostructures with Pt spacer layers up to a thickness of 0.8 nm. This is a consequence of the polycrystalline structure with (111) texture of the Co/Pt multilayers leading to 1–2 nm deep notches generated by neighboring crystallites at the interface. Thus, Pt layers with thicknesses equal or less than 0.8 nm are not continuous and the appearance of the IDW provides an EB field value similar to heterostructures without spacer layer. Thicker Pt layers are continuous and cover the notches between neighboring Co/Pt multilayer crystallites, which allows only indirect magnetic exchange coupling due to the spin polarization in the Pt. Consequently, the EB field becomes strongly reduced. Based on the interface morphology and taking into account the spin polarization in the Pt spacer layer an exchange length of about 0.4 nm was found.

Curriculum Vitae

Personal Information

Surname(s)/First name(s): **Schubert/Christian**
Address(es): Bahnhofstraße 3
09390 Gornsdorf - Saxony - Germany
Telephone(s): +49176 22890774 (private)
Email(s): christian_schubert@ymail.com (work)
christian.schubert@me.com (private)
Nationality(-ies): German
Date of Birth: May 13th, 1984

Personal Statement

Since making the decision to choose Physics as one of my main subjects in the last two years of school, my enthusiasm and interest in my subject has done nothing but deepen. This enthusiasm is that which continually provides me with new motivation to discover phenomena and to grasp complex physical correlations. Furthermore, whilst self-motivation means that I can work effectively on independent projects, I also possess the ability to motivate others and work as part of a team to achieve the best results possible. My independent work also comes hand in hand with a free and creative side which is an asset to any project work. These sides of my personality are fostered through my long standing commitment to my free-time activities, which include running (long-distance and marathon), riding and restoring motorbikes, practicing martial arts and playing the saxophone (both for personal pleasure and in a jazz/funk group).

C. Schubert, *Magnetic Order and Coupling Phenomena*, Springer Theses,
DOI: 10.1007/978-3-319-07106-0,

Spoken Languages

Mother tongue(s)	**German**				
Self-assessment					
European level (*)	**Understanding**		**Speaking**		**Writing**
	Listening	Reading	Spoken interaction	Spoken production	
English	C2 Independent user	C2 Independent user	C1 Independent user	C1 Independent user	C1 Independent user
French	A2 Basic user	A2 Basic user	A1 Basic user	A1 Basic user	A1 Basic user

(*) *Common European Framework of References (CEF) level*

Education

PhD 2009–2013 (Chemnitz University of Technology)	Final grade: **summa cum laude**
Diplom Physik (BA & MA Physics) 2003–2008 (Chemnitz University of Technology)	Final grade (MA): **1.4**; Diploma thesis: 1.3; Final grade (BA): 1.9
Abitur (equivalent to English A-Level) 2002	Primary subjects: Maths 1.0 (equivalent A) & Physics 1.5 (equivalent A)

Qualifications and Awards

2002 Prize from the DPG (German Physics Society)
2002 Jugend forscht Environmental Prize (for an energy saving research project)

Selected Work Experience

Science assistant (Present)	**Chemnitz University of Technology, Germany**
Physics assistant (January 2005–December 2008)	**Chemnitz University of Technology, Germany:** Responsibility for setting up and maintenance of Physics experiments used in student teaching (e.g. beta spectroscopy experiments)
Care assistant–part-time (July 2003–June 2005)	**CURA Retirement Home, Germany:** Responsible for the complete care of up to ten elderly residents.
Care assistant–full-time (August 2002–June 2003)	**CURA Retirement Home, Germany:** My state obligatory civil service was completed in a retirement home, where I had responsibility not only for the general well-being of the residents, but also for their medical requirements.
Builder's assistant (July 2000)	**Loeser Bau, Germany:** General manual labour.
Practical work experience (May 1999)	**KS Elektronik, Germany:** Preparation of circuit boards with electronic devices.

Skills

- Computer literacy in Linux & Microsoft Windows, word processing (e.g. LaTeX, Adobe Acrobat & Microsoft Word), processing of graphics (e.g. Corel Draw), Microsoft Excel & Power Point, data processing (e.g. Origin), software development in C++, Mathematica and internet research
- Conception and assembling of scientific setups (thin film deposition machines and complex measurement devices)
- Good communication and networking skills
- Public speaking, writing & oral presentation of reports.

List of publications

- **2013**

A. Hassdenteufel, B. Hebler, C. Schubert, A. Liebig, M. Teich, M. Helm, M. Aeschlimann, M. Albrecht, and R. Bratschitsch, *Thermally assisted all-optical helicity dependent magnetic switching in amorphous* $Fe_{100-x}Tb_x$ *alloy films*, Advanced Materials, (2013).
C. Schubert, B. Hebler, H. Schletter, A. Liebig, M. Daniel, R. Abrudan, F. Radu, and M. Albrecht, *Interfacial exchange coupling in Fe-Tb/[Co/Pt] heterostructures*, Physical Review B **87**, 054415 (2013).

- **2012**

C. Brombacher, C. Schubert, M. Daniel, A. Liebig, G. Beddies, T. Schumann, W. Skorupa, J. Donges, S. Häberlein, and M. Albrecht, *Chemical ordering of FePt films using millisecond flash-lamp annealing*, Journal of Applied Physics **111**, 1 (2012).
M. Fronk, C. Schubert, F. Haidu, C. Scarlat, K. Dorr, M. Albrecht, D. R. T. Zahn, and G. Salvan, *Characterization of organic thin films on ferromagnetic substrates by spectroscopic ellipsometry and magneto-optical Kerr effect spectroscopy*, IEEE Transactions on Magnetics **48**, 11 (2012).

- **2011**

C. Brombacher, C. Schubert, K. Neupert, M. Kehr, J. Donges, and M. Albrecht, *Influence of annealing time on structural and magnetic properties of rapid thermally annealed FePt films*, Journal of Physics D: Applied Physics **44**, 355001 (2011).
T. Kosub, C. Schubert, H. Schletter, M. Daniel, M. Hietschold, V. Neu, M. Maret, D. Makarov, and M. Albrecht, *Coercivity enhancement in exchange-biased CoO/* $Co_3Pt bilayers$, Journal of Physics D: Applied Physics **44**, 015002 (2011).

- **2010**

D. Makarov, J. Lee, C. Brombacher, C. Schubert, M. Fuger, D. Suess, J. Fidler, and M. Albrecht, *Perpendicular FePt-based exchange-coupled composite media*, Applied Physics Letters **96**, 062501 (2010).

Zeitfracht Medien GmbH
Ferdinand-Jühlke-Straße 7
99095 Erfurt, Deutschland
produktsicherheit@kolibri360.de